AN AMAZONIAN RAIN FOREST

The Structure and Function of a
Nutrient Stressed Ecosystem and the
Impact of Slash-and-Burn Agriculture

MaB

MAN AND THE BIOSPHERE SERIES

Series Editor: J.N.R. Jeffers

VOLUME 2

AN AMAZONIAN RAIN FOREST

The Structure and Function of a
Nutrient Stressed Ecosystem and the
Impact of Slash-and-Burn Agriculture

Edited by
C.F. Jordan

*Institute of Ecology, The University of Georgia,
Athens, Georgia, USA*

PUBLISHED BY

AND

The Parthenon Publishing Group
International Publishers in Science, Technology & Education

Published in 1989 by the United Nations Educational, Scientific and Cultural Organization.
7 Place de Fontenoy, 75700 Paris, France—Unesco ISBN 92-3-102629-1

and

The Parthenon Publishing Group Limited
Casterton Hall, Carnforth,
Lancs LA6 2LA, UK — ISBN 1-85070-230-6

and

The Parthenon Publishing Group Inc.
120 Mill Road,
Park Ridge
New Jersey, NJ, USA — ISBN 0-940813-82-3

Printed and bound in Great Britain by Butler and Tanner Ltd.,
Frome and London

British Library Cataloguing in Publication Data

An Amazonian rain forest: the Structure and function of Nutrient stressed ecosystem
and the impact of slash and burn Agriculture.
  1. Tropical forests.   Soils. Nutrient cycles
  I. Jordon, Carl F.
  631.4'1

ISBN 1-85070-230-6

Library of Congress Cataloging-in-Publication-Data

An Amazonian rain forest: the structure and function of a nutrient stressed ecosystem
  and the impact of slash and burn agriculture/edited by C.F. Jordan.
      p.   cm. — (MAB: Man in the biosphere series: v. 2)
  Bibliography: p.
  Includes index.
  ISBN 0-940813-82-3: $45.00 (U.S.)
    1. Rain forest ecology—Amazon River Region.   2. Shifting cultivation—
  Environmental aspects—Amazon River Region.   3. Forest productivity—Amazon
  River Region.   I. Jordan. C.F.   II. Series: MAB   (Series); 2.
  QH112.A45   1989   89-9316
  634.9'0981'1—dc20   CIP

SERIES PREFACE

Unesco's Man and the Biosphere Programme

Improving scientific understanding of natural and social processes relating to man's interactions with his environment, providing information useful to decision-making on resource use, promoting the conservation of genetic diversity as an integral part of land management, enjoining the efforts of scientists, policymakers and local people in problem-solving ventures, mobilizing resources for field activities, strengthening of regional co-operative frameworks ... These are some of the generic characteristics of Unesco's Man and Biosphere Programme.

Unesco has a long history of concern with environmental matters, dating back to the fledgeling days of the organization. Its first Director-General was biologist Julian Huxley, and among the earliest accomplishments was a collaborative venture with the French Government which led to the creation in 1948 of the International Union for the Conservation of Nature and Natural Resources. About the same time, the Arid Zone Research Programme was launched, and throughout the 1950s and 1960s this programme promoted an integrated approach to natural resources management in the arid and semi-arid regions of the world. There followed a number of other environmental science programmes in such fields as hydrology, marine sciences, earth sciences and the natural heritage, and these continue to provide a solid focus for Unesco's concern with the human environment and its natural resources.

The Man and Biosphere (MAB) Programme was launched by Unesco in the early 1970s. It is a nationally based, international programme of research, training, demonstration and information diffusion. The overall aim is to contribute to efforts for providing the scientific basis and trained personnel needed to deal with problems of rational utilization and conservation of resources and resource systems, and problems of human settlements. MAB emphasizes research for solving problems: it thus involves research by interdisciplinary teams on the interactions between ecological and social systems; field training; and applying a systems approach to understanding the relationships between the natural and human components of development and environmental management.

MAB is a decentralized programme with field projects and training activities in all regions of the world. These are carried out by scientists and technicians from universities, academies of sciences, national research laboratories and other research and development institutions, under the auspices of more than a hundred MAB National Committees. Activities are undertaken in co-operation with a range of international governmental and non-governmental organizations.

Further information on the MAB Programme is contained in *A Practical Guide to MAB*, *Man Belongs to the Earth*, a biennial report, a twice-yearly newsletter *InfoMAB*, MAB technical notes, and various other publications. All are available from the MAB Secretariat in Paris.

Man and the Biosphere Book Series

The Man and the Biosphere Series has been launched with the aim of communicating some of the results generated by the MAB Programme to a wider audience than the existing Unesco series of technical notes and state-of-knowledge reports. The series is aimed primarily at upper level university students, scientists and resource managers, who are not necessarily specialists in ecology. The books will not normally be suitable for undergraduate text books but rather will provide additional resource material in the form of case studies based on primary data collection and written by the researchers involved; global and regional syntheses of comparative research conducted in several sites or countries; and state-of-the-art assessments of knowledge or methodological approaches based on scientific meetings, commissioned reports or panels of experts.

The series will span a range of environmental and natural resource issues. Currently available in press or in preparation are reviews on such topics as control of eutrophication in lakes and reservoirs, sustainable development and environmental management in small islands, reproductive ecology of tropical forest plants, the role of land/inland water ecotones in landscape management and restoration, ecological research and management in alpine regions, rain forest regeneration and management, non-conventional conservation and the role of biosphere reserves in the quest for alternatives, assessment and control of non-point source pollution, research for improved land use in arid northern Kenya, ecological and social effects of large-scale logging of tropical forest in the Gogol Valley (Papua New Guinea), changing land use patterns in the European Alps.

The Editor-in-Chief of the series is John Jeffers, until recently Director of the Institute of Terrestrial Ecology in the United Kingdom, who has been associated with MAB since its inception. He is supported by an Editorial Advisory Board of internationally-renowned scientists from different regions of the world and from different disciplinary backgrounds: E.G. Bonkoungou (Burkina Faso), Gonzalo Halffter (Mexico), Otto Lange (Federal Republic of Germany), Li Wenhau (China), Gilbert Long (France), Ian Noble (Australia), P.S. Ramakrishnan (India), Vladimir Sokolov (USSR) and Anne Whyte (Canada).

A publishing rhythm of three to four books per year is envisaged. Books in the series will be published initially in English, but special arrangements will be sought with different publishers for other language versions on a case-by-case basis.

An Amazonian rain forest

The contents of this book arose from an intensive study of the ecology of the rain forest at a research site in the Amazon Territory of Venezuela, at San Carlos de Rio Negro. The book does not, however, attempt to present the results of the whole of that study, which represented a major research project with international funding and support. It concentrates, instead, on a limited range of questions related to nutrient stress in tropical rain forest, namely:

(1) Are the plants and animals of rain forest under nutrient stress, and, if so, how have they adapted to that stress?

(2) What are the effects of slash and burn agriculture on the nutrient status and productivity of tropical forest sites?

The answers to these questions are of crucial importance for the management and conservation of tropical forest, and it is necessary, therefore, to consider the exent to which the results obtained at San Carlos can be applied to other tropical forests.

The book is a personal assessment by the author of the results that he obtained from his work on the San Carlos project and not every tropical forest ecologist may agree with that assessment. The results are published here so that they can be read, discussed and criticised by other scientists, for that is the way in which science 'works'. Resource managers, administrators, politicians and interested, but non-scientific, members of an increasingly well-informed general public will find much that is of both interest and value in the book, at a time when the rapid disappearance of tropical forest is becoming an issue of international concern. Much of that concern is currently focused on the effects of global climatic warming, but the fate of essential nutrients when tropical forests are felled is also of considerable importance, particularly for the countries in which the forests are situated.

The general reader will find in this book much information which is usually hidden in incomprehensible scientific reports.

The tropical forests of the world represent a rapidly disappearing resource with an astonishing diversity of species and communities of plants and animals. Understanding of the dynamics and ecology of these forest systems is essential if we are to make the right decisions about their conservation or about their exploitation for timber and conversion to agriculture. This book deals with one important aspect of that ecology, ie. the nutrient status of the forests and the fate of those nutrients when the forest is cleared.

CONTENTS

Frontispiece: Shelter belonging to a shifting cultivator on the banks of the Rio Negro near San Carlos, Venezuela.

INTRODUCTION

A. SCIENTIFIC BACKGROUND

In his classic book *The Tropical Rain Forest,* Richards (1952) suggested that nutrient stocks are often low in tropical rain forests, and, that when rain forests are cut and burned for agricultural purposes, nutrient losses result in rapid declines in crop productivity. His description of nutrient dynamics before and after rain forest cutting (p. 219) is as follows:

> In the rain forest climate, as in all climates in which the movement of water in the soil is predominantly downwards, the trend of soil development is always towards impoverishment. Soluble substances are continually being washed down into the deeper layers of the soil and removed in the drainage water. The most important common characteristic of all rain-forest soils, whether of the red earth or the podzol type (except perhaps some very immature soils) is thus their low content of plant nutrients. This being so, it seems paradoxical that rain- forest vegetation should be so luxuriant. The leached and impoverished soils of the wet tropics bear magnificent forest, while the much richer soils of the drier tropical zones bear savanna or much less luxuriant forest. This problem has been considered by Walter and Milne for African forests, and by Hardy for those of the West Indies, and all these authors reach a similar conclusion. In the Rain forest the vegetation itself sets up processes tending to counteract soil impoverishment and under undisturbed conditions there is a closed cycle of plant nutrients. The soil beneath its natural cover thus reaches a state of equilibrium in which its impoverishment, if not actually arrested, proceeds extremely slowly.
>
> Fresh plant nutrients are continually being set free by the decomposition of the parent rock. Provided the horizon in which they are set free is not too deep for the tree-roots to reach, a part of these nutrients is taken up by the vegetation in dilute solution. Some of these substances are fixed in the skeletal material of the plant – the cell walls – others remain dissolved in the cell sap. Eventually all of them are returned to the soil by the death and subsequent decomposition of the plant or its parts. The top layers of the soil are thus being continually enriched in plant nutrients derived ultimately from the deeper layers. The majority of the roots,

1

including nearly all the 'feeding' roots, are in the upper layers of the soil. Most of the nutrients set free from the humus can be taken up again by the vegetation almost immediately and used for further growth. The loss, if any, must be very slight; Milne has shown that in the Usambara Rain forest the electrolyte content of the streams is very low. It can thus be seen that in a mature soil the capital of plant nutrients is mainly locked up in the living vegetation and the humus layer, between which a very nearly closed cycle is set up. The resources of the parent rock are only necessary in order to make good the small losses due to drainage.

The existence of this closed cycle makes it easy to understand why a soil bearing magnificent Rain forest may prove to be far from fertile when the land is cleared and cultivated. When the forest is felled the capital of nutrients is removed or set free in the soil and the humus layer is often destroyed at the same time by burning and exposure to the sun. As Milne says: 'The entire mobile stocks are put into liquidation and as is usual at a forced sale, they go at give-away prices and the advantage reaped is nothing like commensurate with their value'. Crops planted where rain forest has been cleared may thus do very well for a few seasons, benefiting from the temporary enrichment of the soil, but before long, unless special measures are taken, a sterile, uncultivable soil may develop. On the ordinary system of native cultivation practised in rain-forest areas it is rare for more than two or three harvests to be obtained in succession without a long intervening period of 'bush fallow'. Even in British Honduras, where the annual rainfall is not more than about 180 cm, the yield of maize on forest clearings falls from about 350-450 kg in the first year to about 180-270 kg in the third. On the very poor porous rain-forest soils such as the Wallaba podzol in British Guiana it may be impossible to obtain even one crop. The changes in the soil due to the clearing or thinning of the forest emphasize the delicate equilibrium of soil and vegetation in a natural rain forest.

Since the publication of *The Tropical Rain Forest,* there have been several studies of the nutrient cycles in undisturbed tropical forests, for example, El Verde, Puerto Rico (Odum, 1970), Banco, Ivory Coast (Bernhard-Reversat, 1975), Panama (Golley *et al.,* 1975), Pasoh, Malaysia (Kato *et al.,* 1978), as well as studies of nutrient dynamics during slash and burn agriculture, for example, Ghana (Nye and Greenland, 1964), Venezuela (Harris, 1971), Peru (Denevan, 1971; Sanchez *et al.,* 1983), Amazon region (Brinkman and Nascimento, 1973), Thailand (Zinke *et al.,* 1978), and many others. Nevertheless, the question of the importance of nutrients in tropical rain forests and the loss of nutrients following their cutting has remained open to question. For example, the slash and burn studies cited above showed that soil stocks of most nutrients during cultivation were higher than stocks in the undisturbed forest even at the time cultivation was abandoned. These results cast doubts on the idea that nutrient loss is responsible for declining productivity during shifting cultivation. Other criticism of the hypothesis that nutrients are important in controlling structure and function of tropical rain forests has come from Harcombe (1977a,b) and Proctor (1983), who found little evidence for nutrient stress at the sites that they studied.

Questions about the importance of nutrients in a tropical rain forest and in slash and burn agriculture derived from rain forest are addressed in this book. The results are specific to the study site – at San Carlos de Rio Negro in the Amazon Territory of Venezuela – but the forest at the study site is compared with forests in other tropical regions to indicate the extent to which results can be generalised.

The questions addressed in this book are:

(1) Are the plants and animals of the rain forest at San Carlos under nutrient stress?

(2) If they are under nutrient stress, have they adapted to this stress, and if so, how?

(3) How do the forests at San Carlos compare with other tropical forests?

(4) How do nutrient cycles and net primary productivity of the undisturbed forest change due to slash and burn agriculture?

(5) Following abandonment of slash and burn agriculture, is recovery of the forest inhibited by lack of nutrients?

(6) What do the results of this study mean for management of tropical forests?

B. PROJECT BACKGROUND

The studies were part of the 'San Carlos' Project, named after the village near the field site in Venezuela. The project was located at San Carlos because, in the early 1970s, there was a realization that pressures for development of the Amazon Territory in Venezuela were increasing, yet there was very little ecological knowledge about the region with which to plan rational development (Medina *et al.*, 1977). In response to this need, scientists from the Ecology Center at the Instituto Venezolano de Investigaciones Científicas (IVIC), Caracas, began searching for a field station. San Carlos was selected because it was close to a wide variety of ecosystems typical of the upper Rio Negro region, and because it was easily accessible through an air-strip adjacent to the village.

Project studies were coordinated through the Ecology Center, IVIC, Caracas, Venezuela. Dr Ernesto Medina (IVIC) was project director, and the steering committee consisted of Drs. Rafael Herrera (IVIC), Hans

Klinge (Max Planck Institute, FRG) and Carl Jordan (USA). Support for the project was provided in part by the Organization of American States, the UNESCO Man and the Biosphere Program, the United States National Science Foundation (NSF), and the Science Foundation of Venezuela (CONICIT), as well as IVIC, which provided technical, laboratory and office support. The studies emphasized in this book resulted from research ideas proposed to NSF in a series of proposals entitled '*Nutrient Dynamics of a Tropical Rain Forest Ecosystem, and Changes in the Nutrient Cycle due to Cutting and Burning*'.

CHAPTER 1

DESCRIPTION OF THE AMAZON REGION AND THE SAN CARLOS STUDY SITE

A. LOCATION

The village of San Carlos de Rio Negro, latitude 01° 56′ N, longitude 67° 03′ W, is in the Amazon Territory of Venezuela, close to the common boundary point of Venezuela, Colombia and Brazil (Fig. 1.1). It is located about 15 km

Fig. 1.1. Map of northern portion of South America showing location of San Carlos and the Rio Negro.

5

below the confluence of the Casiquiare and the Guainia Rivers which join to form the Rio Negro (Fig. 1.2), the largest southerly flowing tributary of the Amazon. The Casiquiare begins in a bifurcation of the Orinico River at one of the few places in the world where a watershed divide passes through the middle of a river. Because of the bifurcation, it is possible to travel upstream from San Carlos to the Orinoco, and then downstream towards northern Venezuela entirely by boat. This study was carried out on a site 4 km east of the village of San Carlos.

B. CLIMATE

1. Regional climate

In the lowland tropics, the most variable climatic factor is precipitation. In the Amazon Basin, annual totals range from less than 2000 mm to almost

Fig. 1.2.   Map of the southern portion of the Amazon Territory of Venezuela.

4000 mm (Fig. 1.3). Seasonal periodicity of rainfall is greatest in the southern and eastern portion of the basin and generally decreases towards the north and west (Fig. 1.3), although there are local exceptions. San Carlos de Rio Negro is located in a region with high precipitation and low seasonality.

2. Climate at San Carlos

A climate diagram, based on data collected between 1951-58 and 1971-78 at a government meteorology station at San Carlos, is given in Fig. 1.4. During the period of measurement, the average yearly temperature was 26°C, with an average minimum of 22°C, and an average maximum of 31.5°C. The average yearly rainfall was 3565 mm. The average night-time humidity was between 93 and 96 per cent, and day-time humidities averaged between 65 and 70 per cent (Heuveldop, 1980).

C. GEOLOGY

1. Regional

Hundreds of millions of years ago, in the Pre-Cambrian era, the region which now comprises the Amazon Basin consisted of a massive geologic shield. During the Paleozoic era, a depression formed in the middle of this

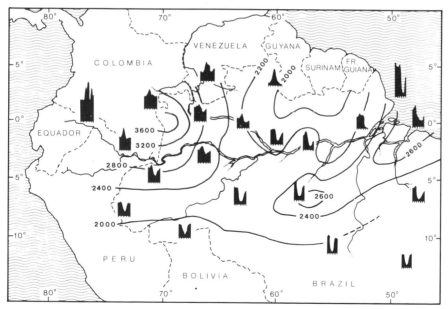

Fig. 1.3. Rainfall patterns in the Amazon Basin (adapted from Salati, 1978). Isohyets are mm per year. Graphs show relative monthly distribution from Jan. to Dec.

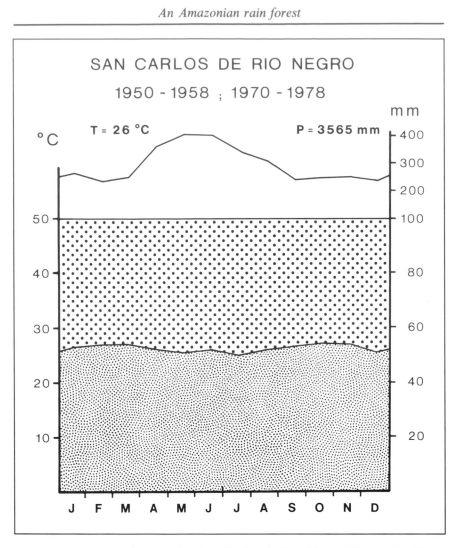

Fig. 1.4. Climate diagram for San Carlos de Rio Negro, Venezuela. (from Heuveldop 1980). The lowermost data line is the average monthly temperature scaled on the left axis. The uppermost line is monthly precipitation scaled on the right axis. The shaded area indicates months with more than 100 mm of precipitation.

shield, leaving to the north what is known today as the Guiana Shield and to the south the Brazilian shield (Putzer, 1984). There has never been a major geological uplift in the central and eastern Amazon Basin, but during the Jurassic, Cretaceous and Tertiary periods, major mountain building in the west formed the Andes and, as a result, westward drainage from the Amazon Region was blocked. The central Amazonian depression became a basin, where sediments eroded from the Guiana and Brazilian Shields were deposited. Those sediments form what is sometimes today called the Amazon Planalto, a relatively flat surface which extends up to 250 meters above sea level in the western end of the basin (Beek and Bramao, 1968). The sediments consist of clay materials that have been severely leached and weathered for 100 million years or more, and consist of minerals such as kaolinite (Kronberg *et al.,* 1979, 1982) which are among the last in the mineral weathering sequence (Jackson *et al.,* 1948). Because of the severe leaching and weathering, the minerals contain low levels of nutrient elements either as part of their crystal structure or adsorbed to clay surfaces.

The leaching of silica from these minerals is the so-called laterization process, and results in iron oxides such as goethite and aluminum oxides such as gibbsite (Brady, 1974). However, formation of a hard laterite crust is relatively rare (Sanchez, 1976), probably because rapid regeneration of vegetation on cleared areas prevents the severe drying necessary for hard-pan formation. Where hard-pan is desired, such as for roads and landing strips in the Amazon region, it is necessary to periodically scrape the areas to prevent re-establishment of vegetation (personal observation). The deep, intense weathering which occurs in the Amazon Basin does result in the formation of deep layers of partially decayed bedrock, or saprolite, which prevents root penetration to the unweathered bedrock below (Jenny, 1980).

2. San Carlos area

The present southwestern limit of the Guiana Shield is about 200 km northeast of San Carlos. However, columnar remnants of very old sandstone formations south of this limit suggest that the Shield was more extensive at one time (Herrera, 1979). These flat-topped mountains are known locally as 'tepuyes'. Isolated dome-shaped granitic batholiths, or 'inselbergs' also occur in the region.

The terrain in the region of San Carlos is gently rolling, with hills up to 40 meters higher than surrounding lowland (Fig. 1.5). The hills and valleys are a reflection of the surface of the underlying granitic bedrock. On the hills, clay derived from the granite extends up to, or close to, the surface (Figs. 1.6 and 1.7), and the depressions between the granitic hills are overlain with sand deposits. Consequently, the entire area gives the impression of a flat sea with scattered islands of higher terrain. The repeating pattern of hills and troughs is associated with a repeating pattern of soils and vegetation. In Fig. 1.5, vegetation on the hills appears as darker patches, while that in the lower areas is lighter in color.

D. SOILS

On top of the hills, where the clay formed from the granite is exposed at or near the soil surface (Fig. 1.7), the soils have been classified as an Oxisol

9

Fig. 1.5. Aerial view of the terrain near San Carlos de Rio Negro, Venezuela. Patches of dark green are caused by distinctive vegetation on Oxisol hills which contrasts with the lighter caatinga vegetation between the hills.

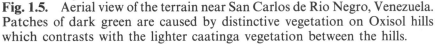

(Dubroeucq and Sanchez, 1981, Appendix Table C.1.1.). In places, lateritic concretions (plinthite) constitute greater than 50 per cent of the material in the upper soil horizons (Herrera, 1979, Figs. 1.8, 1.9). On these Oxisols near San Carlos, roots are concentrated near the surface, and these, together with litter and humus, form a mat (Fig. 1.8) which ranges up to 40cm in thickness. On the sides of hills in some areas near San Carlos, the soil is almost free of plinthite and the clay is a whitish color (Fig. 1.10, Appendix Table C.1.2). This soil has been classified as an Ultisol (Dubroeucq and Sanchez, 1981).

The soils in the areas between the hills are comprised of coarse sands (Table C.1.3). These sands are the well known Podsols or Spodosols of the upper Rio Negro region (Klinge 1967, Figs. 1.11, 1.12).

E. VEGETATION

Because the annual rainfall at San Carlos is between 2000 and 4000mm, the forest would be classified as tropical moist forest, according to the scheme of Holdridge (1967). Since the forest (but not individual trees) is continuously green, Walter (1971) would classify it as continuously wet tropical rain forest. The forest would be included in Richards' (1952) *Tropical Rain-Forest Formation.*

10

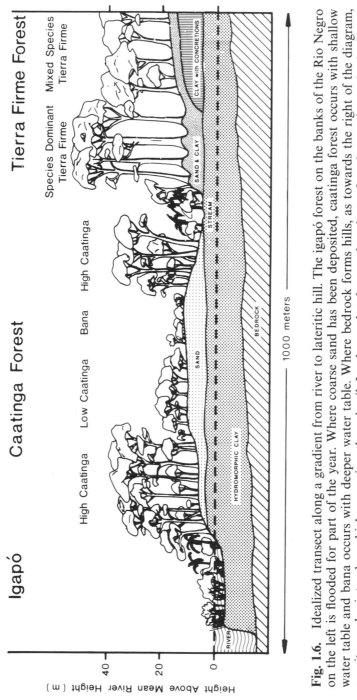

Fig. 1.6. Idealized transect along a gradient from river to lateritic hill. The igapó forest on the left is flooded for part of the year. Where coarse sand has been deposited, caatinga forest occurs with shallow water table and bana occurs with deeper water table. Where bedrock forms hills, as towards the right of the diagram, granite grades into clay which comprises the subsoil. In places, the clay reaches the surface, but at the study site it was covered with a shallow layer of fine sand. On the shoulders of the hills, there sometimes occurs a mixture of fine sand and clay which supports a forest dominated by one or a few species.

11

Fig. 1.7. Saprolite or highly weathered rock formed from underlying granite on top of the hills near San Carlos. Exposure is a result of a road-cut by a bulldozer. For scale, note machete stuck in saprolite, just to the right of picture-center. On top of the saprolite is a remnant cap of soil which supported a forest before clearing.

Fig. 1.8. Profile of top 40 cm of soil on a hill site. Lateritic concretions are visible mixed in with the sand. Note mat of fine roots and humus on soil surface and concentration of larger roots directly below.

Fig. 1.9. Deep profile of soil on a hill site. The penknife marks the transition from the overlaying sand/plinthite horizon to the lower reddish yellow horizon higher in clay.

An idealized transect showing variation in soils as a function of topography (Fig. 1.6) also shows how vegetation is correlated with topography, soil and water table. In the forests on the Oxisol hills (Fig. 1.13), species' diversity is high and there is no strong dominance by any one species. The height and diameter of the trees and biomass of the forest is not particularly large. On the sides of the hills on the Ultisols, the trees are taller and larger in diameter (Fig. 1.14), but diversity is lower, and often just one or a few species are dominant.

The vegetation occurring on the Spodosols depends primarily on the depth of the water table, but nutrients also may be important (Klinge and Herrera, 1978; Herrera, 1979). Where the water table is deepest, a very reduced type of vegetation, locally called 'bana', occurs (Sobrado and Medina, 1980). Although the water table often reaches the surface of bana sites during storms, it quickly drops after rain ceases, because the coarse sand cannot hold much moisture. A few days without rain leaves the bana soils in a droughty condition.

With the decreasing depth of water table along the transect, bana grades into intermediate vegetation, sometimes called campina or low caatinga, and then into high caatinga (Klinge, 1978; Klinge and Medina, 1978). (This can be seen in the background, Fig. 1.15). The caatinga of the Rio Negro region is called 'Amazon caatinga' to distinguish it from the caatinga of the

13

Fig. 1.10. Profile of Ultisol. Markers are every 50 cm. Black color in upper soil horizon is due to organic matter.

arid region of northeastern Brazil (Anderson, 1981). Amazon caatinga forests resemble the so-called heath forests, or Kerangas, of Southeast Asia (Whitmore, 1975).

Seasonally flooded soils along the banks of major streams and rivers support a forest variable in height, called igapo or rebalse, where trees are partially or completely under water for several months each year.

Because of the remoteness of San Carlos and the humid climate of the region, it seemed unlikely that the forests in the area had ever been burned in the recent geological past, either through the action of man or due to natural causes. However, charcoal samples collected from an Oxisol site were shown by radiocarbon dating to be 250 ± 60 years old, and samples from other sites ranged from 250 to 6260 years old (Sanford *et al.,* 1985). Charcoal samples also have been found in soil samples from many other areas near San Carlos (Saldarriaga, 1985), suggesting that the entire region has experienced disturbances in the relatively recent past.

F. DRAINAGE WATER

The dark color of the water in the Rio Negro, as well as in many streams draining the forests near San Carlos, is a striking characteristic of the region. Although the Rio Negro is a 'black water' river (Sioli, 1975), the water is not really black. Alexander von Humboldt (1821) described the color as being 'of an amber color, whenever it is shallow, and of a dark brown like coffee grounds, wherever the depth of the waters is considerable'. The water is clear, as opposed to the main channel of the Amazon which is turbid. The

Fig. 1.11. Profile of Spodosol. The black "B" horizon which marks the accumulation of iron, aluminum, and organic matter is overlain by the coarse leached "A" horizon sands.

15

Fig. 1.12. Close-up of pit in Spodosol. The root mat on top of the picture grades sharply into a lower layer of coarse black humus. The line between the humus and the leached "A" horizon is very sharp. Ground water has accumulated in the bottom of the pit.

Fig. 1.13 Mixed tierra firme forest on Oxisol, taken from the experimental plot after cutting but before burning (see Chapter V.).

Fig. 1.14. Forest on Ultisol near San Carlos.

Fig. 1.15. Bana vegetation on Spodosol. In the background, gradation into caatinga can be seen.

17

dark color of the water in the Rio Negro is due to humic materials dissolved in the water. Janzen (1974) suggested that the high concentrations of humic acids in black water rivers, which are typical of regions low in soil nutrients, is a result of high phenol production by plants as a defense against herbivory. While it is true that the San Carlos area is remarkably free of insects compared to the clear water regions of the upper Orinoco (personal observation), the black color of the water may be due more to low clay content of the Spodosol soils (St. John and Anderson, 1982). In soils with high clay content, complex organic compounds are adsorped on the clay surfaces and are decomposed before they can be leached, but on deep sands the phenols percolate rapidly to drainage streams. We observed that soil water collected from beneath the litter was amber in both the Oxisol and Spodosol sites, but collections from 40 cm deep were clear in the Oxisol, while they remained amber in the Spodosol.

CHAPTER 2

NUTRIENT STRESS, AND THE PLANT AND ANIMAL COMMUNITIES NEAR SAN CARLOS

A major goal of the San Carlos project was to examine the question of whether the forest there was under nutrient stress and, if so, how this stress affected the structure and function of both an undisturbed forest and an agricultural ecosystem derived from the forest.

How can it be determined whether the plants and animals in an ecosystem are under stress? A number of reactions, on the individual, community and ecosystem levels have been suggested as characterizing stress. Woodwell (1970) compared the reaction of a forest community under radiation stress with reactions to other stresses and found that, in general, high stress causes a decrease in photosynthesis, an increase in respiration, a decrease in net productivity, a reduction in structure and biomass, and a decrease in species' diversity.

Odum (1985) has presented a table showing eighteen different trends which may occur in a stressed ecosystem (Table 2.1). Not all trends occur in all stressed ecosystems, and some trends may be limited to certain types of stress. Nevertheless, if several of these trends are observed in an ecosystem, it suggests strongly that the ecosystem is under stress.

Rapport *et al.*, (1985) examined five ecosystem characteristics in reports about twenty-one ecosystems that were under varying types of stress, from intensive harvesting to receiving pollutant discharges. They found that:

1. In six ecosystems, the nutrient pool decreased, but in four it increased.

2. In five ecosystems, the primary productivity decreased, but in three it increased.

3. In all seventeen ecosystems for which data were available, size distribution decreased.

19

4. In all seventeen ecosystems for which data were available, species diversity decreased.

5. All twenty ecosystems for which data were available showed symptoms of 'retrogression', a general term indicating change towards an earlier stage of succession, and characterized by many of the trends listed in Table 2.1.

The characteristics of ecosystems, communities and individuals given by Woodwell (1970), Rapport (1985) and Odum (Table 2.1) are used here as criteria to determine whether the ecosystem at San Carlos may be under stress. Some of these characteristics have been studied as part of the San Carlos project, as follows:

For the producer community:

> Forest biomass
> Net primary productivity
> Species diversity
> Nutrient leaching
> Nutrient cycling index

For the consumer communities, the decomposers and the below-ground community:

> Population density (community structure and biomass)
> Size of organisms
> Species diversity
> Respiration

A. PRIMARY PRODUCERS

1. Biomass

The above- and below-ground biomass of San Carlos caatinga forest and forest on the Oxisol hill (hereafter referred to as the 'Oxisol forest') is compared with the biomass of five other tropical evergreen forests in rows 1 and 2 of Table 2.2. This table, together with Table 2.3, is used throughout this book to compare the change in a variety of ecosystem characteristics as a function of environment, and both tables will be referred to again. Table 2.4 gives the locations of the ecosystems, and Table 2.5 lists the references and footnotes for these tables.

The above-ground biomass of the two San Carlos forests is less than the forests studied in the Ivory Coast, Malaysia, Costa Rica and Panama, but is greater than that for the Puerto Rican forest. The biomass of the Puerto Rican forest may be low because it is a lower montane forest rather than a lowland forest, or because it sustained some hurricane damage several decades before the study. If low biomass is an indicator of stress, the comparative data suggest that the San Carlos forests may be under greater stress than the other tropical forests studied, with the possible exception of the Puerto Rican forest.

The below-ground biomass data show that the caatinga forest has a greater root boimass than any of the other forests in Table 2.1, and the Oxisol forest has a greater root biomass than all the forests except the Puerto Rican forest. Large root biomass may be an adaptation of plants growing on nutrient-poor soil (Chapin, 1980).

The relatively low above-ground biomass and the relatively high below-ground biomass of the forests at San Carlos suggest, then, that both the caatinga forest on Spodosol and the Oxisol forest are under greater stress than many other tropical forests.

Table 2.1 Trends expected in stressed ecosystems (from Odum, 1985).

Energetics

1. Community respiration increases
2. Photosynthesis/respiration ratio becomes unbalanced (usually to <1.0, but in exceptional cases, >1.0).
3. Photosynthesis/biomass and respiration/biomass ratios increase.
4. Importance of auxiliary energy increases.
5. Exported or unused primary production increases

Nutrient cycling

6. Nutrient turnover increases.
7. Cycling index decreases (less efficient recycling).
8   Nutrient loss increases.

Community structure

9. Proportion of r-strategists increases.
10. Size of organisms decreases.
11. Lifespans of organisms or parts (leaves for example) decrease.
12. Food chains shorten.
13. Species diversity decreases and dominance increases; if original diversity is low, the reverse may occur.

General system-level trends

14. Ecosystem becomes more open (i.e. input and output evnironments become more important as internal cycling is reduced).
15. Autogenic successional trends reverse (Succession reverts to earlier stages).
16. Efficiency of resource use decreases.
17. Parasitism and other negative interactions increase, and mutualism and other positive interactions decrease.
18. Functional properties (such as community metabolism) are more resistant to stress than are species composition and other structural properties.

Table 2.2 Ecosystem characteristics of seven tropical moist and rain forests. Numbered references and footnotes are given in Table 2.5 (see page 25 for details of the references)

Parameter	Amazon Caatinga San Carlos, Venezuela	Oxisol forest San Carlos, Venezuela	Lower montane rain forest El Verde, Puerto Rico	Evergreen Forest Banco, Ivory Coast	Dipterocarp forest Pasoh, Malaysia	Lowland rain forest La Selva, Costa Rica	Moist forest, Panama
1. Root biomass (t/ha)	132[1]	56[2]	72.3[7]	49[17]	20.5[19]	14.4[14]	11.2[23]
2. Aboveground biomass (t/ha)	268[1]	264[2]	228[8]	513[17]	475[21]	382[15]	326[23,25]
3. Root/shoot ratio	.49	.21	.32	.10	.04	.04	.03
4. Root distribution, % in superficial root mat	26[1]	20[2]	~0[7]	~0	~0[19]	~0[14]	—
5. Specific leaf area (cm^2/gm)	47[3]	65[4]	61[7]	—	88[21]	139[26]	131-187[23,25]
6. Leaf area index	5.1[1]	6.4[4]	6.6[9]	—	7.3[21]	—	10.6-22.4[23,25]
7. Predicted leaf biomass (t/ha)	10.8	9.8	10.8	—	8.3	—	10.4
8. Leaf litter production (t/ha/yr)	4.95[2]	5.87[2]	5.47[8]	8.19[17]	6.30[22]	7.83[16]	11.3[23]
9. Predicted turnover time of leaves (row 7/row 8) (years)	2.2	1.7	2.0	—	1.3	—	0.9
10. Aboveground wood productivity (t/ha/yr)	3.93[2]	4.93[2]	4.86[8]	4.0[17]	6.4[21]	—	—
11. Leaf decomposition (k)	0.76[2]	0.52[2]	2.74[10]	3.3[17]	3.3[22]	3.47[16]	3.2[23]
12. Biomass/phosphorus ratio, leaf litter fall	2631[1]	7237[10]	5000[6]	1365[18]	3282[20]	2024[26]	1319[23]
13. Biomass/nitrogen ratio, leaf litter fall	135[1]	95[2]	—	64[18]	82[20]	52[27]	—

Table 2.3 Nutrient characteristics of the soils at the ecosystem comparison sites. (see page 25 for details of the references)

Parameter	Amazon Caatirga San Carlos, Venezuela	Oxisol forest San Carlos, Venezuela	Lower montane rain forest El Verde, Puerto Rico	Evergreen Forest Banco, Ivory Coast	Dipterocarp forest Pasoh, Malaysia	Lowland rain forest La Selva, Costa Rica	Moist forest, Panama
1. Exchangeable calcium in A_1 soil horizon (meg/100 g)	0.57[1]	0.03[2]	1-5[5]	.09[18]	0.13[19,24]	1.28[13]	36[23]
2. Total calcium in soil (kg/ha)	195[1]	7[2]	176[12]	—	115[19,24]	6530[16]	22,166[23]
3. Soil pH	4.0[1]	3.9[2]	4.3-4.8[11]	4.1[18]	4.3-4.8[23]	4.0[13]	4.7-5.9[28]
4. Total nitrogen in soil (kg/ha)	785[1]	1697[2]	—	6500[18]	6752[19,24]	20,000[16]	—
5. Total phosphorus in soil (kg/ha)	36[1]	243[2]	—	600[18]	44[19,24]	7000[16]	22[23]

Table 2.4 Locations of the tropical and subtropical forests used in ecosystem comparisons. References are listed in Table 2.5 except for (*) which are from Meentemeyer (undated) and (+) from Luvall (1984). (see page 25 for details of the references)

	Latitude and Longitude	Location	Precipitation mm/yr	Approximate meters above sea level	Average temperature °C	Actual Evapotranspiration mm/yr
San Carlos, Venezuela**	01°56'N 67°03'W	Near the common border of Venezuela, Colombia, and Brazil	3565	100	26	1778
El Verde, Puerto Rico	18°19'N 65°45'W	Luquillo mountains, eastern Puerto Rico	2920	500	23	1752
Banco Forest, Ivory Coast	5°N 4°W	Near Abidjan	2095	100	26	1530*
Pasoh, Malaysia	2°58'N 102°17'E	140 km southeast of Kuala Lumpur	2054	100	26	1515*
La Selva, Costa Rica	10°N 84°W	Foot of central mountains, northeast Costa Rica	4300	40	24+	2153+
Darien Province Panama	8°38'N 78°08'W	Sante Fe	2000	<250	25	1442*

**Two ecosystem sites were near San Carlos.

Table 2.5 References and footnotes for Tables 2.2–2.4.

1. Klinge and Herrera, 1978; Herrera, 1979
2. This book
3. Medina *et al.*, 1978
4. Jordan and Uhl, 1978
5. Jordan and Herrera, 1981
6. Luse, 1970
7. Odum, 1970
8. Jordan, 1971
9. Jordan, 1969
10. Wiegert, 1970
11. Edmisten, 1970
12. Jordan *et al.*, 1972
13. Bourgeois *et al.*, 1972
14. Raich, 1980
15. Werner, unpub. data
16. Gessel *et al.*, 1977
17. Lemee, undated
18. Bernhard-Reversat, 1975
19. Yoda, 1978
20. Lim, 1978
21. Kato *et al.*, 1978
22. Ogawa, 1978
23. Golley *et al.*, 1975
24. Average of 7 plots
25. Average of 2 sites
26. Luvall and Parker, unpub. data
27. Cole and Johnson, undated
28. Golley, unpub. ms.

2. Net primary productivity

It is difficult to judge the influence of stress on net primary productivity from item 2 (photosynthesis/respiration ratio) in Table 2.1 because the ratio could alter either from a change in photosynthesis or in respiration. Further, the short-term response of the ratio could be different to the long-term response. For example, if an ecosystem is subjected to stress by toxic pollutants, photosynthesis will decrease, respiration will increase and the ratio will decrease. However, in an ecosystem which has been subjected to a stress for a long time, the organisms may adapt and respiration may eventually decrease – if not within one generation, then perhaps after several – and low rates of photosynthesis will be masked by low rates of respiration when stress is judged by the ratio.

Since nutrient-stress in the Amazon rain forests, if it exists, is likely to have emerged gradually, it is perhaps better to examine net primary productivity separately, as did Rapport *et al.* (1985).

Net primary productivity is broken down into stems plus branches, roots, and fine litter in Table 2.6, and two of the categories are compared with data from other tropical forests in Table 2.2, parameters 8 and 10. Fine litter (mainly leaves) productivity is lowest in the San Carlos caatinga forest, and the Oxisol forest is third from the lowest. Comparisons of net primary productivity of leaf litter suggest, as did comparisons of biomass, that the forests at San Carlos are under stress. Comparisons of wood productivity (Table 2.2, item 10) is more difficult, because data are not available for all ecosystems.

3. Species' diversity

Decrease in species' diversity is another symptom of stress listed in Table 2.1. Comparisons of species' diversity are difficult because of the variety of indices used, and because the index frequently is influenced by the size of the sample, which often varies between studies. For the purposes of this chapter it was most convenient simply to compare the number of tree species per hectare.

There were found to be 83 species of trees with diameters greater than 10 cm on the one hectare of forest on Oxisol used as the control site for the experiment described in Chapter V. (A list of all terrestrial vascular plants near San Carlos is given in Appendix A, Table A.1). Uhl and Murphy, (1981) compared the diversity of the forest on the Oxisol plot to plots of similar size in other neotropical rain forests and found little difference. Diversity in African forests also appears similar. Richards, (1952) cites studies from Africa which show that, for trees greater than 10 cm in diameter, in a 1.5 ha plot in Nigeria there were 70 species, in a 1.4 ha plot in the Ivory Coast there were 74 species, and in a 1.0 ha plot in Mauritius there were 52

Table 2.6 Summary of net biomass production on Oxisol and Spodosol sites. Periods of measurement indicated in parenthesis. Data are from Chapters V & VI.

Components	Oxisol t/ha/yr	Spodosol t/ha/yr
Stems and branches (1975 - 1983)	4.93	3.93
Roots (1976 - 1978)	2.01	N.D.
Litter (Leaves and twigs < 1 cm, (1975 - 1980)	5.87	4.95
Σ	12.81	

species. In contrast to South America and Africa, the diversity of trees in southeastern Asia appears higher where many sites have more than 100 trees greater than 30 cm diameter per hectare (Whitmore, 1984).

When diversity is used as an index of stress, the forest community on Oxisol at San Carlos exhibits no symptoms of stress, at least compared with limited data from African and other South American sites. One possible explanation is that when a community has existed under stress for many generations, certain symptoms, such as, perhaps, low diversity, disappear due to evolutionary adaptation. In any case, it is not, of course, necessary for stressed communities or ecosystems to exhibit all the symptoms of stress listed in Table 2.1

Table 2.1 does suggest that when initial diversity is low, stress may result in a high diversity. This situation probably occurs more frequently in ecosystems, such as monocultures of agricultural crops, where stress results in invasion by weeds.

4. Nutrient leaching

Nutrient leaching and the net balance between nutrient input and output may be indicators of stress, but these may be more a characteristic of the entire ecosystem rather than just the plant community. However, since the forest trees are the principal agents which recycle nutrients and prevent leaching, these parameters are more conveniently considered here.

A low rate of nutrient leaching and a positive net balance between input and output suggests that the forest may be in an aggrading state as during secondary succession (Vitousek and Reiners, 1975). A high loss rate and a net negative balance suggests that the forest may be senescent, or under stress.

The net balances of nutrients at the Oxisol site were determined by measuring atmospheric deposition and loss through leaching and details of the methods used are given in Appendix C.7. The results (Table 2.7) show that for all ionic species except ammonium and nitrate nitrogen, the input-output balance is positive. The negative balance for nitrogen in water fluxes may be compensated for by gaseous inputs, and the positive balance of the other nutrients may reflect temporal fluctuations in nutrient balance (Figure 2.1) rather than long term accumulation due to aggradation.

To compare leaching and net nutrient balance at San Carlos with these parameters at other forest sites, calcium data were used because of data availability. The comparison (Table 2.8) shows that at San Carlos leaching is lower and the balance is positive and higher than at most other sites. The nutrient leaching and net balance data, then, give no evidence that the San Carlos forests are under stress. However, just as with species' diversity, lack of evidence of stress does not mean that stress has not occurred. It may be that since the stress has occurred over a long period of time, species have had

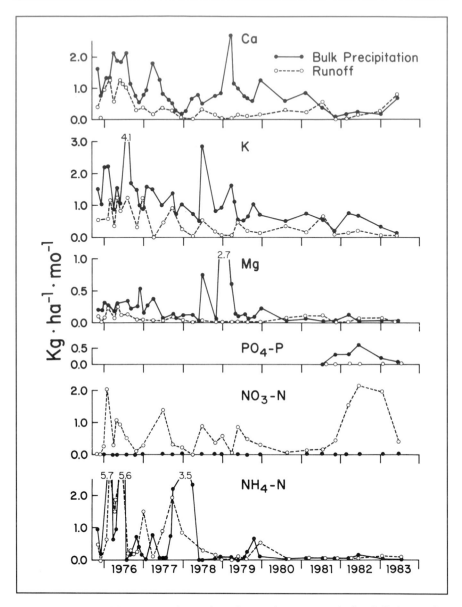

Fig. 2.1. Monthly rates of nutrient input by wet and dry-fall from the atmosphere, and loss by leaching, for the tierra firme 'control' forest on Oxisol (Chapter V, and Appendix C.7).

28

time to adapt in some ways and that this particular symptom of stress is no longer apparent. Some of the possible adaptations will be examined in the next chapter.

5. Nutrient recycling

The 'cycling index' (Table 2.1, item 7) is an indicator of the tightness of a forest's nutrient cycle (Finn, 1976), and, like leaching losses and net nutrient balance, it is an ecosystem level characteristic. It differs from net balance and leaching in that it also considers the internal nutrient cycle of the forest, along with the input and output. A forest with a tight internal cycle has a high cycling index, whereas one with a leaky cycle has a low index, and this may be indicative of stress.

One way of defining the cycling index of an ecosystem is to divide the amount of material recycling through the ecosystem by the total amount of that material moving straight through it (Finn, 1976). However, a problem with this definition is that the index depends in part, on the what constitutes the 'straight through' pathway, that is, it depends on the structure of the model used to calculate the index.

The cycling index can also be calculated as that proportion of the total amount of material entering a compartment which is recycled and eventually returns to that compartment (Finn, 1978). For example, suppose that, annually, 10 kg per hectare of a nutrient enter the soil by throughfall,

Table 2.7 Average input of nutrients into the control plot via wet fall and dry fall and loss due to leaching from soil. Data, except for phosphorus, is averaged over a 7.79 year period from September 15, 1975 to June 30, 1983, during which time there was 27,100 mm of precipitation and 13,250 mm of runoff. Data for phosphorus is for July 31, 1981 to June 30, 1983 during which time there was 6789 mm of precipitation and 3441 mm of runoff.

	Input kg/ha/yr	Input concentration* ppm, volume weighted	Leaching losses kg/ha/yr	Leachate concentration ppm, volume weighted	Net yearly change kg/ha/yr
Calcium	8.83	0.25	3.48	0.20	+5.35
Potassium	10.59	0.30	3.66	0.22	+6.93
Magnesium	2.44	0.07	0.96	0.06	+1.48
NH_4-N	5.39	0.15	7.28	0.43	−1.89
NO_3-N	0.74	0.02	4.28	0.25	−3.54
PO_4-P	0.25	0.01	0.0	0.00	+0.25

*Concentration is of nutrients in the water in rainfall collectors. It would include nutrients in dry-fall that have been washed in the collectors.

29

Table 2.8 Runoff (R), atmospheric input (A), and difference (A–R) for calcium in various ecosystems arranged in increasing order of calcium runoff.

Formation or Association, and Location	Calcium $(kg.ha^{-1}.yr^{-1})$ Runoff (R)	Input (A)	A-R	Author
1. Tropical rain forest, Malaysia	2.1	14.0	+11.9	Kenworthy 1971
2. Rain forest, Amazon Basin Spodosol	2.8	16.0	+13.2	Herrera 1979
3. Rain forest, Amazon, Basin Oxisol	3.5	8.8	+5.3	Jordan, this report
4. Evergreen forest, Ivory Coast	3.8	1.9	–1.9	Bernhard-Reversat 1975
5. Pine forest, North Carolina	4.1	6.5	+2.4	Swank & Douglass 1977 Johnson & Swank 1973
6. Douglas fir forest, Washington	4.5	2.8	–1.7	Cole *et al*, 1967
7. Coniferous forest, northern Minnesota	4.5	3.1	–1.4	Wright 1976
8. Mixed coniferous forest, New Mexico	4.9	7.6	+2.7	Gosz 1975
9. Hardwood forest, North Carolina	6.9	6.2	–0.7	Swank & Douglass 1977 Johnson & Swank 1973
10. Oak-pine forest, Long Island	9.7	3.3	–6.4	Woodwell & Whittaker 1967 Woodwell *et al*, 1975
11. Beech forest on sandstone, Germany	12.7	12.8	+0.1	Heinrichs & Mayer 1977
12. Spruce forest on sandstone, Germany	13.5	12.8	–0.7	Heinrichs & Mayer
13. Northern hardboard forest, New Hampshire	13.9	2.2	–11.7	Likens *et al*, 1977
14. Tropical rain forest, New Guinea	24.8	0	–24.8	Turvey 1974
15. Aspen Forest, Michigan	19.4 to 38.8	8.3	–11.1 to –30.5	Richardson & Lund 1975
16. Mixed mesophytic forest, Tennessee	27.4	10.5	–16.9	Shugart *et al*, 1976
17. Montane tropical rain forest, Puerto Rico	43.1	21.8	–21.3	Jordan *et al*, 1972
18. Tropical moist forest, Panama	163.2	29.3	–133.9	Golley *et al*, 1975

stemflow, litter-fall and tree-fall, that 2 kg are leached from the soil, and that the stock in the soil remains constant. This means that 8 kg per hectare are taken up each year by the roots of trees and eventually return again to the soil. The cycling index for the soil would be 0.8.

The cycling index of three nutrient cations was calculated for the superficial mat of roots and humus, the mineral soil and a combination of these two compartments for the forest on the Oxisol site at San Carlos. The conceptual models on which the indices are based are shown in Figs. 2.2-2.4, and the methods used to quantify the models are given in Appendix C.

The cycling indices are calcuated in Table 2.9. The rate of nutrient movement into the humus mat compartment on the forest floor is the sum of throughfall, stemflow, litter-fall and tree-fall (Table 2.9, row a). The rate of nutrient uptake from the mat (row c) is the difference between row a and the total leached from the mat (row b). Since all the nutrients taken up by the vegetation eventually return again to the humus, row c divided by row a is the cycling index for the humus compartment (row d).

The same procedure is used to calculate the cycling index for the mineral soil alone. The difference between the rates of nutrient movement into and

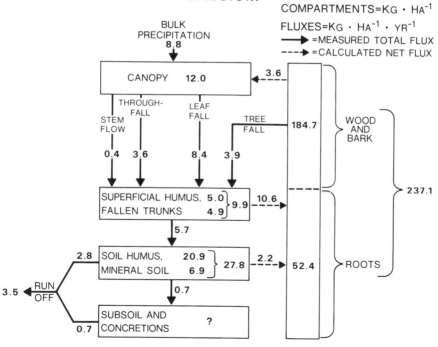

Fig. 2.2. Model of the calcium cycle on the Oxisol site at San Carlos.

out of the mineral soil (row f) is the rate of nutrient uptake from soil to vegetation. Row f divided by the amount entering the mineral soil (row b) is the cycling index for the mineral soil (row g).

The cycling index for the humus mat plus the mineral soil (row h) is calculated as the total recycling rates (rows c and f) divided by the total entering the forest floor (row a).

The data in Table 2.9 quantify the proportion of recycled nutrients that are cycled directly in the San Carlos forest, when 'direct nutrient cycling' is defined as the movement of nutrients from decomposing litter to roots without passage through the mineral soil. Total cycling is the amount taken up by the roots from both the humus mat (row c) and the mineral soil (row f). The proportion of total recycled nutrients which is recycled 'directly' (row i) is the uptake from the humus (row c) divided by the total uptake. Direct nutrient cycling clearly is important for nutrient cations in the Oxisol forest at San Carlos.

Comparisons of the cycling index for San Carlos with the index for forests in other regions is difficult because of the lack of comparable data. Although watershed studies often include nutrient cycling data (Likens *et al.*, 1977), nutrient runoff as determined in watershed studies may not be suitable for

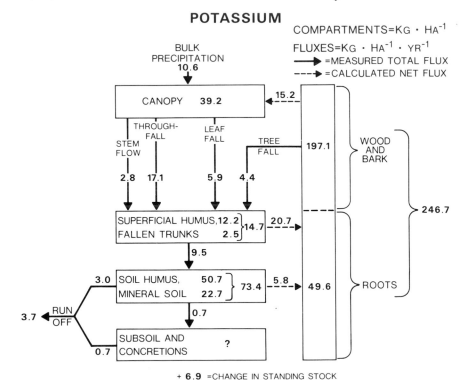

Fig. 2.3. Model of the potassium cycle on the Oxisol site at San Carlos.

the calculation of a recycling index because an unknown proportion of the nutrients in the runoff could be moving directly from weathered rock to drainage streams, without cycling through the vegetation.

Table 2.10 lists three sites besides the San Carlos Oxisol plot for which enough data are available for comparisons. One is the lower montane forest in Puerto Rico, the second is a deciduous, mixed mesophytic forest on calcium-rich soil in Tennessee and the third is a Douglas fir forest in Washington. The cycling index of calcium in the Puerto Rican forest is much lower than that for San Carlos, while in the Tennessee forest it is only slightly lower. A low cycling index for calcium may be related to the high soil calcium content at these sites (Jordan and Herrera, 1981). With greater availability in the soil, highly efficient calcium recycling is not required for the survival of the forest. Although the Douglas fir forest in Washington has a high soil calcium content, its calcium cyling index is almost as high as that of San Carlos, which may reflect the successional state of the Washington forest. If the forest is not close to steady state biomass, the yearly increase in biomass could account for the large proportion of calcium entering the soil which is taken up by the roots rather than lost through leaching.

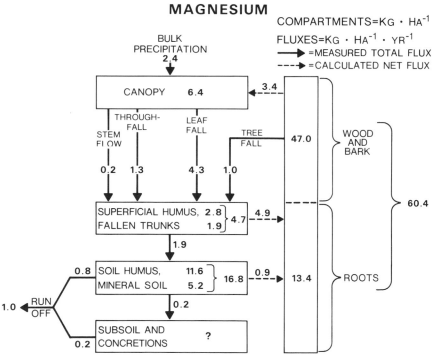

Fig. 2.4. Model of the magnesium cycle on the Oxisol site at San Carlos.

The cycling index for potassium at all three sites was higher than for calcium, which could indicate a high demand for that nutrient. A comparison of the cycling indices for calcium and potassium suggests that potassium may be more critical than calcium in the Oxisol forest at San Carlos.

The comparison of the cycling indices between sites indicates that any stress which may occur at the San Carlos site does not measurably affect the nutrient cycling index, as determined by comparisons with other ecosystems for which data are available. Lack of evidence of stress could mean either that stress does not occur at the San Carlos site, or, that nutrient recycling mechanisms have evolved to compensate. Because some of the other characteristics of stress listed in Table 2.1 do occur at San Carlos, it is more likely that the lack of effect on the nutrient cycling index results from evolutionary adaptations of the species to the stress. The nature of the adaptations is the subject of the next chapter.

B. CONSUMERS

In contrast to the studies of the forest trees at San Carlos, most of the studies of consumers were orientated towards species or specific taxonomic groups. The parameters compared here to detect evidence of stress in species or populations of consumers at San Carlos are population density, density of

Table 2.9 Cycling indices for cations in the humus mat and total soil compartments in the Oxisol site. Data from Figs. 6.1–6.3.

		Ca	K	Mg
a.	Total entering forest floor*	16.3	30.2	6.8
b.	Total leached through mat*	5.7	9.5	1.9
c.	Total cycled from mat (a-b)*	10.6	20.7	4.9
d.	Cycling index. mat compartment $\left(\dfrac{c}{a}\right)$	.65	.69	.72
e.	Total leached out of soil*	3.5	3.7	1.0
f.	Total cycled from soil (b-e)*	2.2	5.8	0.9
g.	Cycling index, soil alone $\left(\dfrac{f}{b}\right)$	.39	.61	.47
h.	Cycling index, mat + soil $\left(\dfrac{c+f}{a}\right)$ or $\left(\dfrac{a-e}{a}\right)$	.79	.88	.85
i.	Proportion of total cycling that occurs "directly" $\left(\dfrac{c}{c+f}\right)$ or $\left(\dfrac{c}{a-e}\right)$	.83	.78	.84

*Rates in kg/ha/yr

34

Table 2.10 Comparison of cycling indices of calcium and potassium.

	a. Amount entering forest floor (litter fall + through fall) kg/ha/yr	b. Leaching from rooting zone kg/ha/yr	c. Cycling Index $\left(\dfrac{a-b}{a}\right)$
Calcium			
San Carlos Oxisol (this report)	16.3	3.5	.79
Montane forest Puerto Rico (Jordan et al. 1972)*	62.9	43.1	.31
Mesophytic forest on calcium rich soil Tennessee (Shugart et al. 1976)	84.1	27.4	.67
Douglas fir forest Washington (Cole et al. 1967)	18.5	4.5	.76
Potassium			
San Carlos	30.2	3.7	.88
Puerto Rico	138.9	20.8	.85
Washington	15.8	1.0	.94

*Tree-fall data lacking

colonies or nests, size of individuals, species' diversity and activity such as the rate of microbial transformations. Not all parameters were measured in every study, but studies where at least one parameter was determined are included.

1. Small mammals

A small mammal trapping study was carried out over a four-hectare area for a one-year period. The methods used are given in Appendix B.1, and a list of species in Appendix A.2. A comparison of population densities at San Carlos with those in other areas (Table 2.11) shows that the small mammal density at San Carlos is comparable to that of a tropical forest in Malaya, but lower than other tropical forests and temperate forests, suggesting that the San Carlos community is, or has been, under stress. The population densities of forest mammals seem to be determined by the potential resources, and ultimately energy, available to them in specified habitats (Robinson and Redford, 1986).

2. Birds

Studies of the bird fauna were limited to observations between 17-22 December, 1980 by Francois Vuilleumier. Despite the abbreviated visit, the following quotation from his unpublished field notes are pertinent to the question of density and diversity of birds at San Carlos.

> My overwhelming impression of the avifauna around San Carlos was its poverty in both number of families and species and in numbers of individuals of each species. This remark applies to each habitat visited, whether tierra firme forest, caatinga alta, caatinga baja, secondary woodlands, riverine forest, or other habitats. I made both day and night visits to several habitats, including the two types of caatinga, trying to include in the survey both diurnal and nocturnal (or crepuscular) birds, yet was surprised not to encounter (either to see or to hear) several species belonging to widespread tropical groups, which I had expected to find.
>
> Sixty-six species of birds were detected (Appendix A.3), 56 of which could be identified with certainty. A similar period of observation in a similar variety of habitats in other parts of Amazonia should yield approximately 80-120 species. Numbers in this range have been observed in Costa Rica (Orians, 1969) and western Ecuador (Vuilleumier, 1978).

3. Fish

a. Population density

One of the few practical measures of fish population density, and therefore one of the most common, is the weight of the fish captured per unit of fishing

Table 2.11 Densities of small mammal communities in forests from various localities. Table prepared by R.F.C. Smith.

Number of species in community	Number of individuals per ha	Habitat and Location	Authority
17 (10 rodents: 7 marsupials)	5.1	Tropical forest, Venezuela	This study
16 (11 rodents: 5 marsupials)	22.4	Dry tropical forest, Panama	Fleming (1975)
13 (9 rodents: 4 marsupials)	13.4	Moist tropical forest, Panama	Fleming (1975)
11 (all rodents)	4.9	Primary forest, Malaya	Harrison (1969)
5 (4 rodents: 1 tree shrew)	7.1	Secondary forest, Malaya	Harrison (1969)
4 (3 rodents: 1 insectivore)	14.8	Hardwoods, South Carolina	Smith *et al.* (1974)
6 (5 rodents: 1 insectivore)	14.6	Pine and hardwoods, North Carolina	Nabholz (1973)
15 (12 rodents: 3 insectivores)	54.8	4 Deciduous forests, West Virginia	Kirkland (1977)
12 (8 rodents: 4 insectivores)	42.0	3 Coniferous forests, West Virginia	Kirkland (1977)
6 (all rodents)	53.0	Creosote bush, New Mexico	Whitford (1976)

effort. Monthly catch per unit effort (kilos of raw fish per man hour of labor) by all fishing methods was calculated for the period April, 1979 to April, 1981 by K. Clark. Work time includes the hours spent traveling, fishing, building traps and collecting bait, but does not include the value of boats, motors, gasoline or fishing gear. Further details are given in Appendix B.2., and the results compared with data from other studies in Table 2.12.

When compared to fish catch rates elsewhere in lowland South America, the San Carlos fishery is very poor. The catch per unit effort is similar to that achieved by the relatively unacculturated Indian groups. However, San Carlos fishermen utilize technology (e.g. outboard motors, headlamps, batteries) often unavailable to less acculturated Indians. Bari Indians, for instance, catch 0.42 kilos of fish per man hour without steel hooks or other introduced technology (Beckerman, 1980). The San Carlos catch is only one-third that of fishermen at Itacoatiara, Brazil, who use similar small-scale fishing gear such as trotlines, handlines and spears (Smith, 1979).

About two-thirds of Rio Negro fishes are small, maturing at 50g or, frequently, much less. Moreover, a number of species, particularly large predatory fish, appear to be dwarfed in the upper Rio Negro, failing to reach the proportions reportedly attained elsewhere (K. Clark, unpublished.).

b. Species diversity

Although fish production in the upper Rio Negro is poor, the richness of the region's icthyofauna is remarkably high, with approximately 300 species having been collected within a 50 km radius of San Carlos (families given in Appendix Table A.4.). In comparison, the Mississippi River drainage (about 15 per cent of the area of North America) has some 250 species and there are fewer than 200 freshwater species in all of Europe (Lowe-McConnell, 1975; Roberts, 1973). Despite the great richness of the Amazon fish fauna, few major groups are represented. In the upper Rio Negro about 40 per cent of the fishes are characoids, 35 siluroids (catfishes) and 7 per cent gymnotoids (knife fishes).

While there is considerable overlap between the icthyofaunas of the upper Rio Negro and those of the Orinoco and the central Amazon, a number of large species of considerable economic importance elsewhere in the Amazon region are notably absent from the San Carlos area. Among these are the pirarucu and aruana (Osteoglossidae), the bocachicos of the family Prochilodontidae, the large serrasalmids of the genus *Colossoma,* and a variety of large pimelodid catfishes.

4. Arthropods

Arthropods were collected from soil samples (soil and litter extraction), litter and humus layer (pitfall traps), understory (sweep samples) and canopy (Malaise traps) in forest plots on both the Oxisol and Spodosol sites. The methods are described further in Golley (1977).

Table 2.12 Fish catch per unit of effort (kilos uncleaned fish per man hour) in lowland South America.

Location	Source	Fish Catch (kg/hr)	Comments
R. Madeira, Brazil	Goulding 1981	3.2 – 6.7	1974-77 commercial catch reported as 25.7–53.9 kg per man day
Itacoatiara, Brazil (Central Amazon)	Smith 1979	4.4	1977 total commercial catch
Itacoatiara, Brazil (Central Amazon)	Smith 1979	1.9	1977 commercial catch by small scale methods similar to those used in San Carlos
R. Sao Lourenco, Brazil	Werner *et al.* 1979	0.76*	Bororo Indians – subsistence fishing
San Carlos de Rio Negro, Venezuela		0.62	This study
R. das Mortes, Brazil	Werner *et al.* 1979	0.45*	Xavante Indians – subsistence fishing
Catatumbo region, Colombia	Beckerman 1980	0.42	Bari Indians – subsistence fishing (*Prochilodus* fishery only)
Pará, Brazil	Werner *et al.* 1979	0.22*	Mekranoti Indians – subsistence fishing
Orinoco region, Venezuela	Lizot 1977	0.08–0.17	Yanomami Indians – subsistence fishing
Maranhao, Brazil	Werner *et al.* 1979	0.06*	Kanela Indians – subsistence fishing

* Reported as weight of cleaned fish, converted to weight of uncleaned fish assuming an 11.6 per cent mean weight loss in cleaning as determined at San Carlos (Clark, unpubl. data).

a. Population density

In the soil at both sites in San Carlos there were about 0.3 individuals per cm^2 of soil surface as determined by soil extractions. In a southern hardwood forest in the U.S.A., Gist and Crossley (1975) found about three individuals per cm^2, and densities in the soil-litter fauna for the U.S.A. IBP forest biome studies ranged up to 13.7 animals per cm^2 (McBrayer *et al.*, 1977). Clearly the Amazonian soil population densities are lower than those of temperate hardwood forests.

Over a five-day period the pitfall traps on the Oxisol plot at San Carlos yielded 23 individuals and those on the Spodosol plot yielded 32 indiviuals. Calculations based on the data of Gist and Crossley (1975) indicated that in a southern hardwood forest about 18 arthropods would be caught per trap if the same sampling intensity had been used (Golley, 1977). In Puerto Rico less than two individuals were captured per pitfall trap over a twelve-day period (McMahan and Sollins, 1970), whereas, in Panama, Duever and Child (unpublished) found between 9 and 164 animals per trap over a three-day period.

In the understory, five samples of 100 sweeps per sample were taken with a beating net at each of both sites. In the Oxisol site, 55 individuals were captured, and in the Spodosol, 75. By comparison, in a second growth forest in the northern part of Venezuela, 3210 individuals were captured with 800 sweeps (Janzen *et al.*, 1976), in Costa Rica, Janzen and Schoener (1968) collected between 448 and 4857 individuals in samples of 2000 sweeps and, in Panama, Duever and Child (unpublished) collected between 640 and 2060 individuals per sample of 100 sweeps.

Thus the data gathered suggests that the population density of arthropods at San Carlos is relatively low in the soil and understory but in the root-humus layer studied by pitfall traps it is comparable to other areas.

b. Diversity

The number of morphotypes per individual (calculated as the total number of morphotypes in sample divided by number of indiviuals in sample) is only a rough estimate of species' diversity, but it is the only index available to compare the diversity of species at San Carlos with other regions. The number of morphotypes per individual in the San Carlos sweep net samples (Table 2.13) is higher than the species/individual ratios calculated from the Venezuelan and Costa Rican data of Janzen *et al.* (1976) and Janzen and Schoener (1968), which range from 0.08 to 0.41 species per individual. The diversity for the canopy insects, as determined from the Malaise traps, is also relatively high.

Although the insect density data then suggest that some insect communities may be under stress, the diversity data that is available shows no evidence for this.

Table 2.13 Relation of morphotypes to individuals captured by four sampling methods in two Amazonian forests (from Golley, 1977).

	Spodosol			Oxisol		
	Morphotypes	*Individuals*	*Morpho/Ind*	*Morphotypes*	*Individuals*	*Morpho/Ind*
Malaise Traps	35	86	0.41	45	104	0.43
Sweep Nets	45	75	0.60	45	55	0.75
Pit Traps	57	634	0.09	57	465	0.12
Soil Extraction	22	150	0.15	23	235	0.10

c. Density of specific taxa

Two other studies of arthropods were carried out as part of the San Carlos project. They are relevant here because they also shed light on the question of stress and population density at San Carlos compared to other regions. In a study of leaf cutter ants, Haines (1983) made a complete search of 1.2 ha of forest floor on the Oxisol site and 0.3 ha on the Spodosol site. A comparison of colony density at San Carlos with data from other sites (Table 2.14) shows that the San Carlos region is very low. Although there are nests of *Atta* sp. near San Carlos, none were encountered in the study plots. Three nests of *Acromyrex octospinosus* Reich. were found in the Spodosol plot, but data for comparison with other regions is lacking. The biomass of leaf cutter ants at San Carlos, as indicated by the density of nests, suggests that there may be a stress which is affecting this community.

The dry weight termite biomass on three plots at San Carlos was determined and converted to rates of litter consumption (Table 2.15, Salick *et al.*, 1983). It was found that litter consumption on the Oxisol site was greater than on the Spodosol sites, and was comparable to the $55.5 g/m^2/yr$ calculated for a Malaysian rain forest (Matsumoto, 1976). This value, whilst exceeding some African savanna estimates (e.g. $6.2 g/m^2/yr$, I.C.I.P.E., 1980), does not approach values reported for other African savannas (e.g. $178.6 g/m^2/yr$, Wood and Sands., 1978).

Table 2.14 Densities of *Atta* colonies in three tropical forests.

Forest Location	Forest Age	Atta Species	Density Colonies/ha	Number of Colonies Found
San Carlos de Rio Negro, Venezuela (Haines 1983)	Mature (Oxisol and Spodosol sites)	*Atta cepalotes* L.	0	0
Isthmus of Panama (Haines 1978)				
Pipeline road	42	*Atta colombica* Guerin	0.75	21
		Atta cephalotes L.	0.07	2
		Both species together	0.82	23
Barro Colorado	100-200	*Atta colombica* Guerin	0.002	1
		Atta cephalotes L.	0.019	7
		Both species together	0.02	8
La Selva, Costa Rica (Haines, unpub.)	Secondary forest	*Atta* species	3.75	6
	Primary forest	*Atta* species	1.9	6

In contrast to the data for leaf cutter ants, the termite data from San Carlos, at least on the Oxisol site, is therefore not strongly suggestive of stress.

5. Earthworms

Earthworms in the San Carlos forests were studied by Nemeth and Herrera (1982) and their density and biomass compared with values for other tropical ecosystems (Table 2.16). Although the biomass reported for San Carlos should be regarded as an underestimate because it does not include the giant genus *Andiodrilus* it nevertheless shows that values for San Carlos are in the same range as, or greater than, values for other tropical forests, even though the values for savannas appear to be greater. Thus the earthworm data gives no evidence of stress in the San Carlos forests.

6. Decomposers

The rates of litter decomposition are a convenient index of the activity, and possibly of the biomass, of all the organisms which are feeding upon the litter. It is an index to the activity of a trophic group, rather than of a taxonomic group. The decay coefficient k (Olson, 1963) has been frequently used in the past as a measure of the rate of decomposition, and the determination of k assumes a single constant exponential decay rate of litter. Although

Table 2.15 Comparison of termite consumption and community parameters for Oxisol, caatinga, and bana sites. Consumption is calculated from dry weight using formulae in Wood and Sands (1978). Per cent total litter consumed is derived from consumption and total forest litter and tree fall per year (Jordan 1978a). Number of termite species and diversity excludes *Anoploterms* spp. and *Nasutitermes* spp. because taxonomic revision is needed. Termite diversity $= -\sum p \ln p$ where p is taken to be the number of plots in which the ith species is found (n) divided $\sum n$ over all species at a site. From Salick *et al.* (1983).

	Site		
	Oxisol	Caatinga on Spodosol	Bana on Spodosol
Termite consumption (g/m^2/yr)	59 ± 80	21 ± 75	32 ± 92
Per cent total litter consumed	5	3	—
Number of termite species	12	10	5
Per cent of termites feeding on organic matter in soils	24	2.3	1.9
Termite diversity	2.15	1.89	1.45

it is now known that rates vary considerably throughout the year, and throughout the period of decomposition of a particular piece of litter, (Swift *et al.,* 1979), nevertheless *k* remains a convenient index for comparing decomposition in various ecosystems. The methods used for determining *k* values at San Carlos are given in Appendix B.3., which shows the average value for the Oxisol site as 0.52 and the Spodosol as 0.76.

Swift *et al.* (1979) present average values of *k* which range from 0.03 in the tundra to 6.0 in some tropical rain forests. In Table 2.2, row 11, *k* values for a series of individual sites are given, and it would appear from this that a *k* of 6.0 may be unusually high. Nevertheless, the comparisons in Table 2.2 show that the leaf litter decomposition rates at San Carlos are lower than for many other tropical forests and much lower than would be expected based on climate, suggesting that the nutrient-poor environment at the site has a strong effect on decomposers.

7. Below-ground community

Soil respiration is an index of the activity of all the below-ground organisms in an ecosystem. It was determined at San Carlos by the technique in which CO_2 absorption in KOH is calculated (Medina *et al.,* 1980), and, using this method, the average soil respiration rates were found to be 114 ± 24 mg $CO_2/m^2/hr$ in the Oxisol and 165 ± 30 mg $CO_2/m^2/hr$ in the Spodosol by Medina *et al.* (1980). Furthermore, by separating the root mat from the soil,

Table 2.16 Density and biomass of earthworms in different tropical and subtropical ecosystems.

Ecosystem	Location	No./m^2	g/m^2	Reference
Savanna	South Africa	74	96	Ljungström and Reinecke (1969)
Tropical Forest	Nigeria	33	10	Madge (1969)
Savanna	Lamto, Ivory Coast	91–400	13.4–54.4	Lavelle (1978)
Gallery forest	Lamto, Ivory Coast	74.7	3.4	Lavelle (1978)
Flooded forest	Uganda	7.4	0.23–3.64	Block and Banage (1968)
Rain forest	Amazonas, Venezuela	32.7–68.4	8.7–16.6	This study (Nemeth & Herrera, 1982)
Tropical pastures	Laguna Verde, Mexico	700	47	Lavelle, Maury and Serrano, 1981
Tropical forest	Laguna Verde	132	9.8	Lavelle, Maury and Serrano, 1981

they were able to estimate that roots contributed between 67 and 86 per cent of total soil respiration.

These results were then compared with soil respiration in seven different ecosystems under different climates in Venezuela (Fig. 2.5). With the exception of one montane site, the soil respiration at sites other than San Carlos (shown by dots) were well correlated with average soil temperature. In contrast, however, the average rates on the San Carlos Spodosol site (upper circle) and on the Oxisol site (lower circle) fell well below the rates that would be predicted based on soil temperature, indicating that the respiration of soil organisms at San Carlos would seem to be strongly stressed by factors other than climate.

The biomass of bacterial and fungal carbon was estimated from initial respiration rates of soil samples (Anderson and Domsch, 1978) and selective inhibition techniques. The total microbial carbon in both the Oxisol and

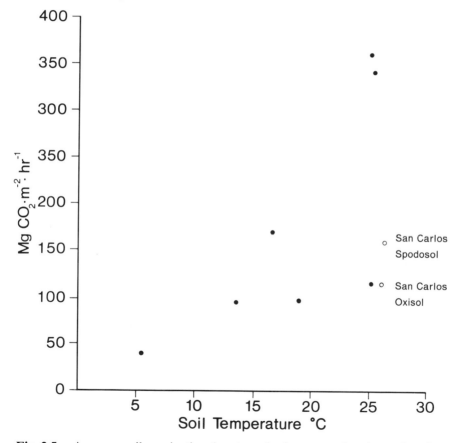

Fig. 2.5. Average soil respiration (roots and micro-organisms) as a function of temperature in ecosystems under different climates in Venezuela. Solid dots are from data of Medina and Zelwer (1972). Circles are from San Carlos.

Spodosol was close to 80mg per 100 grams of soil, with fungi comprising about 75 per cent of the mass in both soils (Caskey, unpublished data). These total values are in the low range of microbial biomass reported for a variety of sites by Anderson and Domsch (1975), who estimated that agricultural and grassland plots from the temperate zone had between 15 and 240mg microbial carbon biomass per 100 grams of soil. The low quantity of decomposers at San Carlos may be part of the reason for the low rates of soil respiration.

8. Nitrogen transformers

The rates of nitrogen fixation and denitrification and the balance between inorganic nitrogen inputs and outputs of the ecosystem are a result of the activity of a wide variety of organisms, including bacteria, algae and lichens. Nitrogen dynamics were measured in both the Spodosol and Oxisol sites at San Carlos (Appendix C.7) but comparisons of results with other ecosystems for which data are available (Table 2.17) is difficult because of very limited data. Nitrogen fixation, which is carried out primarily in the root-humus layer above the soil, is relatively high at San Carlos, whereas denitrification is relatively low, probably because of low numbers of denitrifying bacteria (Gamble *et al.*, 1977). Studies of nitrification by Jordan *et al.* (1979) also revealed almost no activity, which they attributed to the inhibition by secondary plant products leached out of the decomposing litter.

C. NATURE OF THE STRESS AT SAN CARLOS

Some of the characteristics of the ecosystem and the plant and animal communities at San Carlos have suggested that they are under stress, for example, the study of nitrifying bacteria suggested that allelopathic substances may inhibit the activity of that particular community. However, such community-level responses may have evolved as a response to an overall ecosystem level stress. What is the nature of that stress?

The Richards' quotation in the Introduction suggested that, in general, a high potential for nutrient loss may be common in tropical rain forests. Sioli (1975) has argued that the low nutrient content of many of the rivers in the Amazon Basin, and the Rio Negro in particular, reflect a nutrient scarcity in the soils of much of the Amazon. Sanchez (1981) stated that phosphorus deficiency occurs in 90 per cent of the soil of the Amazon region.

It is difficult to avoid a circular argument when discussing the causes and results of stress in natural communities. For example, low nutrient levels cause stress, and stress is indicated by low nutrient levels. Evidence for a relatively low level of a resource in a given community does not necessarily

Table 2.17 Nitrogen comparisons in undisturbed forests.

	NH$_4$-N in Precipitation	NO$_3$ in Precipitation	N Fixation	NH$_4$-N Leached	NO$_3$-N Leached	Denitrification
			$kg.ha^{-1}.yr^{-1}$			
Amazon Rain Forest Tierra Firme on Oxisol	5.4	0.7	16.2[1]	8.4[1]	5.7[1]	2.9[1]
Amazon Rain Forest Caatinga on Spodosol	5.4	0.7	>35[2]	9*[1]	*	—
Seasonal Forest Banco I, Ivory Coast[3]	21.2*	*	—	21.2*	*	—
Hardwood Forest Coweeta, North Carolina[4]	2.7	3.6	12.0	0.6	0.1	10-18
Hardwood Forest Hubbard Brook, New Hampshire[5]	6.5	*	14.2	4.0*	*	—
Douglas Fir Andrew Forest, Oregon[6]	2.0*	*	2.8	1.5*	*	—

[1] Jordan et al. 1982
[2] Herrera and Jordan 1981
[3] Bernhard-Reversat 1975
[4] Todd, Meyer, and Waide 1978; Swank and Dcuglass 1975; Henderson et al. 1978
[5] Borman et al. 1977
[6] Sollins et al. 1980
*Symbol means that values for NH$_4$ and NO$_3$ were combined for precipitation and leaching in original report

mean that that community is under stress – the species may have adapted to the low level and be able to function so that little evidence of stress is apparent. Indeed, in certain ways the communities in the forest at San Carlos appear to be well adapted to exist in an environment where nutrients are generally lower than in forests of other regions.

To avoid the circularity of the stress/adaptation argument in natural communities, nutrient stress can be defined as occurring when there is evidence that one or more nutrients are limiting primary productivity. Vitousek (1982, 1984) has suggested that, in leaves of forest trees, high nutrient-use efficiency for a particular nutrient may indicate that the lack of that nutrient is a factor limiting productivity of the forests. High nutrient-use efficiency for a particular nutrient means that a relatively large amount of biomass is produced with a relatively low amount of that nutrient. Nutrient-use efficiency for the synthesis of a particular tissue is simply the reciprocal of the nutrient concentration in that tissue. If, for example, the concentration of phosphorus in a leaf is very low compared with leaves from other trees, it means that the leaf biomass has been synthesized very efficiently with regard to phosphorus.

Comparisons of nutrient-use efficiency from leaf litter for a wide variety of temperate and tropical forests (Fig. 2.6) show that nitrogen may frequently be a limiting factor in temperate forests. The high nitrogen-use efficiency in the tropical ecosystems are on mountain or white-sand sites, but nitrogen is otherwise generally not limiting in lowland tropical forests (Vitousek, 1984), whereas phosphorus often is. Included in the original figures published by Vitousek are data from the forests on Oxisol and Spodosol at San Carlos. In Fig. 2.6, data from the Oxisol is represented by a circle, and data from the Spodosol by a triangle. Nutrient-use efficiency is relatively high for nitrogen in the Spodosol forest, and very high for both phosphorus and calcium in the Oxisol.

High nutrient-use efficiency in leaves of the forest at San Carlos suggests that a low level of nutrients is the cause of stress in many of the communities and populations in the ecosystem. Because only some of the characteristics of stressed communities are apparent in this ecosystem, it seems that species have adapted in some ways, but not in others, to the nutrient stress.

Other studies at San Carlos also suggest that the low levels of nutrients are limiting the growth of the forest. Curvas and Medina (1988), working on an Oxisol, found that roots of native species fertilized with phosphate had an average growth of $6.43 \, g/m^2/day$, with calcium a growth of $6.75 \, g/m^2/day$, and with ammonium nitrogen a growth of $4.78 \, g/m^2/day$, compared to unfertilized control roots which grew at $3.06 \, g/m^2/day$. Sanford (1987) found strong increases in root growth upwards on stems, when stem flow was enriched with nutrients.

An absence of response to fertilization could have meant that the forest is not now under stress, but it would not have ruled out the possibility that it had been under stress in the evolutionary past. The significance of stress in this case might only become apparent were the forest to be cut down, and agricultural techniques appropriate to nutrient-rich soils attempted in the nutrient-poor soils remaining.

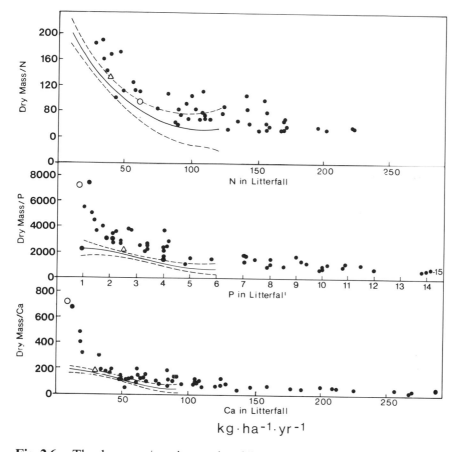

Fig. 2.6. The dry mass/nutrient ratio of litterfall plotted against the amount of that nutrient in litterfall. Each point represents a tropical forest. The circles are the Oxisol site, and the triangles are the Spodosol site at San Carlos. The solid lines represent polynomial regressions (with 95% confidence limits) from a similar analysis of temperate and boreal ecosystems. Fresh leaf litter data was used rather than data from living leaves, since nutrient withdrawal from leaves into twigs before leaf shedding further accentuates the effect of a limiting nutrient. (adapted from Vitousek 1984).

D. CONCLUSIONS

This chapter began with the question: 'Are the plant and animal communities of the rain forests at San Carlos under nutrient stress?'.

It has been shown that certain characteristics of the ecosystem and its communities at San Carlos were indicative of stress. The forest has a relatively low biomass, and productivity at least of leaf litter is also low when compared with other tropical forests. Population densities of small mammals, birds, fish and herbivorous arthropods at San Carlos were also relatively low compared with other regions both in the tropics and in temperate zones, suggesting that these groups at San Carlos may be under stress.

Some populations living or feeding in the humus-root mat layer above the soil surface also showed signs of stress, indicated by low rates of decomposition, soil respiration, nitrification and denitrification. Conversely, other groups in the same layer did not show any marked symptoms of stress, notably arthropods collected in pitfall traps, termites, and earthworms. The relatively high density of these groups which feed in the humus-root layer suggests that this is a stressful environment only for some groups. Other populations, which are adapted to take advantage of an environment high in coarse litter, seem to have found this layer to be hospitable.

In general, however, the structure, biomass and productivity of the populations at San Carlos were indicative of a stressed ecosystem, whereas, in contrast, the species' diversity comparisons gave no indication of stress – except for birds, which were not extensively sampled. Species' diversities for all groups at San Carlos (except birds) were not markedly lower than for other tropical regions, and change in species' diversity may be a reaction to stress characteristic only in the short term. Long term stress, such as nutrient scarcity in a rain forest, may not result in a low species' diversity. However, a discussion of the evolutionary interplay between stress and species' diversity is outside the scope of this book.

In common with the studies of species' diversity, the comparisons of nutrient leaching, net nutrient balance and cycling index also did not indicate stress at San Carlos. However, for these latter three indices, studies had been carried out which showed why this was the case. A major part of the San Carlos project was concerned with studies of nutrient-conserving mechanisms and how they resulted in the efficient recycling of nutrients and the prevention of leaching. An examination of these mechanisms is the subject of the next chapter.

CHAPTER 3

NUTRIENT CONSERVING
MECHANISMS OF THE FOREST

Comparisons of data from studies at San Carlos with studies from other regions suggested that many of the communities in the San Carlos rain forest show some symptoms of stress. Relatively small structure was one symptom exhibited by almost all communities – in the forest tree community by low biomass per hectare, and in the animal communities, by low population density.

In contrast, other characteristics which have been suggested as symptomatic of stress, such as low species' diversity, high nutrient losses and low nutrient cycling index, were lacking. The exhibition of some symptoms of stress, but not others, suggests that species have been able to adapt better in some ways than in others to the stress. The relatively low rates of nutrient loss and the high nutrient recycling index of the San Carlos forest indicates that species of the forest community have evolved well-developed mechanisms to conserve nutrients. Nutrient conservation appears to be particularly critical in the humid tropics, because the year-round high temperatures and rainfall result in a high potential for nutrient losses through leaching of nutrient cations, acidification of the soil, fixation of phosphorus and gaseous loss of nitrogen (Jordan, 1985). The mechanisms for nutrient conservation was a subject of special interest in the San Carlos project, and studies which illustrate these mechanisms are reported in the following sections.

A. ROOTS

1. Biomass

Large root biomass and a high root/shoot ratio is often a characteristic of trees adapted to nutrient-poor soils (Hermann, 1977; Chapin, 1980). A large root biomass occupies more fully the volume of soil where nutrients are held after release from decomposition, and thus the probability of capture of nutrients by roots is greater. Root biomass in the Spodosol forest at San Carlos (Table 2.2, row 1,) is higher than any of the other forests compared in this table, and the root biomass of the Oxisol forest is higher than any except that of the Puerto Rican montane forest.

The root/shoot ratio may be a better indicator of nutrient stress than absolute values of root biomass. Large trees on fertile soil and small, nutrient-stressed trees on infertile soil both may have a large root biomass, but nutrient-stressed trees will have a higher ratio than trees on fertile soil. Table 2.2, row 3, shows that the root/shoot ratio is higher for the San Carlos forests than for any other except, again, for the Puerto Rican forest.

2. Root distribution

Root concentration near the soil surface, or on top of the surface as in the forests of San Carlos (Figs. 1.8, 1.12,) is advantageous to trees that must compete with decomposers or other trees for scarce nutrients. Roots near or on top of the soil surface are intimately mixed with the litter and decomposer organisms and nutrients released from the litter can be taken up almost immediately by surface roots, thereby giving a competitive advantage to trees with such rooting systems.

The concentration of roots near the surface may be advantageous to the entire forest community, as well as to the individual trees. Such concentration, in a thin but complete layer, may be more effective in preventing loss than roots dispersed over a larger volume of soil. The effectiveness of the superficial root mat in intercepting soluble nutrients was demonstrated by Stark and Jordan (1978), who used a sprinkling can to apply phosphorus-32 to the surface of the root mat in the Oxisol site. One 0.1 per cent of the phosporous-32 applied leached down through the root mat to the mineral soil below, while 99.9 per cent was retained in the mat of organic matter and roots. Initial retention was probably through adsorption on the surface of humus, fine litter, soil microorganisms and roots but, later, uptake also was important. The retention of phosphorus in the humus-root mat is important for the continued availability of this nutrient. Once it moves down into the mineral soil, it is rapidly fixed by iron and aluminium compounds and beccomes relatively unavailable to plants (Sanchez, 1976).

Similar experiments with calcium-45, a longer lived isotope, showed that after three months concentrations of this isotope were four or more times higher in roots than in the leaves touching the roots, showing how roots are able to concentrate nutrients (Jordan and Stark, 1978).

In the ecosystem comparisons given in row 4, Table 2.2, above-ground root mats were conspicuous only at the San Carlos sites. Although data on below-ground distribution were not available for the other tropical sites given in this Table, comparisons of distributions in nutrient-rich and nutrient-poor soils of Europe show a much higher concentration of roots near the surface in the nutrient-poor soils (Meyer and Gottsche, 1971).

B. MYCORRHIZAE

Although mycorrhizal associations occur with many species of higher plants, and in rich as well as in poor soils, the mycorrhizal association seems to benefit higher plants most in nutrient-poor soils. Mycorrhizae increase the surface area available to roots for nutrient uptake, and organic compounds released into the soil by mycorrhizae may play a role in solubilizing phosphate, thereby making it more available for uptake (Graustein *et al.,* 1977; Sollins *et al.,* 1981).

Ectomycorrhizae are very common in the nutrient-poor heath forests of the tropics (Singer and Silva Araujo, 1979), such as the caatinga on Spodosol soil at San Carlos. In other rain forest types, trees often form associations with the vesicular-arbuscular endomycorrhizae (St. John, 1980). In a study of the occurrence of mycorrhizae in the Oxisol and Spodosol sites at San Carlos, St. John and Uhl (1983) found that one hundred per cent of the species where determinations could be made were infected by mycorrhizae (Table 3.1).

Mycorrhizae may provide a 'direct' pathway for nutrients moving from decomposing litter to roots (Went and Stark, 1968a,b). The importance of direct cycling for nutrient conservation is that the nutrients move from litter to roots without being exchanged onto mineral soil, where they are more susceptible to loss. The mechanism for loss depends on the nutrient – monovalent cations such as potassium are readily susceptible to leaching losses, phosphorus is bound in the clay, and nitrogen and sulfur may be volatilized.

Evidence for this direct movement of nutrients from litter to roots through ectomycorrhizae was obtained from experiments with phosphorus-32 labeled litter at the Oxisol site near San Carlos by Herrera *et al.* (1978). They isolated the litter in a petri dish on the forest floor, inserted a rootlet through an arm in the side of the dish, sealed the opening and allowed the root to grow for eight weeks. An electron micro-autoradiograph of a cross section through the leaf, the root and the mycorrhizal hyphae connecting them showed radioactivity in all three tissues.

This evidence for direct cycling has been criticized because it has been interpreted by some as suggesting that mycorrhizae were the only, or at least the most important mechanism of nutrient movement from litter to roots. This suggestion was not intended. Nutrients can move from litter through arthropods, algae, lichens, bacteria, earthworms and other forest-floor dwellers as well as mycorrhizae and eventually to the roots. The exact pathway within the surficial layer of litter, humus and roots is not critical

Table 3.1 Mycorrhizal infection as percentage of length of non-woody roots and as percentage of all roots less than 2 mm diameter for common tierra firme and caatinga forest trees (from St. John and Uhl 1983).

Forest Species	% infection[1] ± S.D.	% roots < 2 mm diameter in "infectable" catetory	% infection roots < 2 mm diameter	Number of collections examined	Type of mycorrhizae	% of total aboveground forest biomass
			Tierra Firme Forest			
Licania cf. *heteromorpha* Benth. Chrysobalanacae	100	36.8	36.8	5	VAM	13.9
Caryocar sp. Caryocaracae	78.0 ± 4.35	41.6	32.4	5	VAM	5.6
"Macure" (Probable *Ormosia* sp.) Leguminosae	47.7 ± 6.60	14.7	7.01	5	ECM	4.4
Ocotea sp. Lauracae	87.0 ± 3.97	42.4	36.9	5	VAM	1.3

(continued)

Table 3.1 (*continued*)

Protium sp. Bursuraceae	77.5 ± 4.14	58.0	45.0	5	VAM	0.5
Lecythis sp. Lecythidaceae	unquantifiable low – ~ 25%	27.7	—	5	VAM	0.2
Duguetia sp. Annonaceae	40.0 ± 4.28	41.3	16.5	5	VAM	0.1
"Cabari" (unidentified) Leguminosae	71.3 ± 3.63	32.3	23.0	5	VAM; some ECM	0.1
Average for tierra firme	71.6	38.6	28.2			
Caatinga Forest						
Micrandra sprucei (Muell. Arg.) Schultes Euphorbiaceae	unquantifiable	47.9	—	5	VAM	46.5
Eperua leucantha Benth. Leguminosae	100	30.3	30.3	5	VAM	15.9
Eperua purpurea Benth. Caesalpinaceae	100	31.4	31.4	5	VAM; some ECM	2.7
Manilkara sp. Sapotaceae	100	45.1	45.1	4	VAM	?
Average for caatinga	100	35.6	35.6			

[1]Percentage infection refers to the percentage of the total line intercepts having infected roots. Checks were made only on the fraction of roots < 2 mm diameter judged as infectable.

from the aspect of nutrient conservation. The important point is that following release from the litter, the nutrients are rapidly incorporated into the biomass of living organisms where they are not as readily susceptible to loss as when exchanged on the surfaces of mineral soil. As long as the forest is not disturbed, the nutrients are released gradually from the micro-organisms at a rate which is compatible with the ability of the roots of higher plants to take them up. If direct cycling is defined as movement from decomposing litter through the forest-floor communities without the nutrients entering mineral soil, this is perhaps more meaningful from a nutrient conservation perspective.

The theory of direct cycling also has been criticized because it has been interpreted by some as suggesting that mycorrhizae actually play a role in the decomposition of litter, a suggestion for which little evidence was available. If 'direct' is interpreted as movement from litter to root without passing through mineral soil, rather than as movement entirely through mycorrhizae, the criticism becomes moot. It is worth noting however, that there is recent evidence that ectomycorrhizal fungi may in fact play a direct role in litter decomposition (Janos, 1983; St. John and Coleman, 1983).

C. LEAVES

Evergreen, scleromorphic leaves such as those in the 'bana' type of caatinga forest at San Carlos (Sobrado and Medina, 1980) are tough, long-lived and insect resistant, and often occur in nutrient-poor tropical forests (Medina and Klinge, 1983). Leaves that are tough are usually thick and have a low specific leaf area, which is the area to weight ratio. The specific leaf area in the Spodosol forest at San Carlos is lower than any of the other forests compared in Table 2.2, row 5, and that of the Oxisol forest is lower than other forests except, yet again, the Puerto Rican montane forest. Such leaves may be advantageous to plants where leaf-replacement is expensive in terms of the energy required to obtain replacement nutrients. Leaves that are tough, and consequently long-lived, reduce the necessity for nutrient uptake to replace leaves that are lost (Chapin, 1980). The predicted turnover time of leaves – that is, their lifespan – is relatively long in the San Carlos forests (Table 2.2, row 9).

Thick leaves are also probably more resistant to leaching loss, since nutrient leaching is a function of the area available for leaching to occur. The concentrations of three nutrient cations in throughfall at the San Carlos forests were much lower than the average concentrations determined for a large number of sites throughout the world (Table 3.2).

Secondary plant compounds, such as phenols, may serve as chemical defenses against pathogens and herbivores (Levin, 1976), although in some studies leaf toughness was better correlated with insect resistance than concentrations of phenols in leaves (Coley, 1983; Lowman and Box, 1983). The caloric values of leaves may be an index of the concentrations of secondary plant compounds present. The average caloric content of phenols is 7.6 calories/mg, of alkaloids is 7.8 calories/mg, while cellulose, lignin and glucose are 4.2, 6.3 and 3.7 calories/mg respectively (Domalski, 1972; Lieth, 1975). The caloric values of leaves from the Oxisol forest averaged 5.3 calories/mg, while those from the Spodosol averaged 5.8 calories/mg (Table 3.3). In contrast, leaves of most tropical forests have values of 4.0 or less (Golley, 1961, 1969; Jordan, 1971). This suggests that leaves of the forests at San Carlos may be high in secondary plant compounds, and the inhibition of nitrification in the soils there may be caused by secondary plant compounds leached from the leaf litter (Jordan *et al.*, 1979).

The translocation of nutrients from leaf to twig before leaf abscission is another mechanism by which trees conserve nutrients (Charley and Richards, 1983). A comparison of nutrient concentrations in mature leaves picked from trees and freshly fallen litter from the same trees at three San Carlos sites showed a large reduction in nutrient content of leaves before they fall (Table 3.4). Some of the reduction, especially of potassium, could have resulted from leaching losses. However, because the concentrations of nutrients in throughfall were often equal to or lower than concentrations in rainfall (Jordan *et al.*, 1980, it is likely that nutrient withdrawal was chiefly responsible for the decrease in content.

Epiphylls, the algae, lichens, liverworts and bacteria that live on the surfaces of leaves may also play a role in nutrient accumulation and conservation in tropical forests. Some of these epiphylls fix nitrogen (Forman, 1975), and there is evidence that some of this nitrogen can be translocated from the epiphylls to the host leaf (Bentley and Carpenter, 1984). Epipylls also adsorb nutrients which come in contact with their surface, either dissolved in precipitation (Witkamp, 1970), or suspended in the air (Parker, 1983). Adsorption by epiphylls may have been the cause of the decrease in nutrient concentration in the rain as it passed through the canopy of the forests at San Carlos (Jordan *et al.*, 1980).

D. EFFECTIVENESS OF NUTRIENT CONSERVING MECHANISMS

The comparison of nutrient leaching rates and nutrient cycling indices at San Carlos with values obtained from other sites (Chap. II) suggests that the nutrient conserving mechanisms described here are relatively effective in recycling nutrients and preventing their losses. The effectiveness in maintaining a high recycling index (Table 2.10) is especially noteworthy in view of the higher potential for nutrient loss at San Carlos, where the annual rainfall is higher than at the other sites in this Table (1.2 times higher than at the Puerto Rican site, 2.8 times the Tennessee site, and 2.6 times the Washington site). Another important difference in nutrient loss potential between San Carlos and the temperate zone sites in Table 2.10 is that, due to continual warm temperatures, soil respiration and the production of carbonic acid continue year-round at San Carlos. The dissociation of carbonic acid results in strong leaching potential where concentrations of this acid are high (Jordan, 1985). Despite the higher nutrient-loss potential at San Carlos, the actual loss appears lower than at many other sites, probably because of the highly developed nutrient-conserving mechanisms there.

A very important aspect of these nutrient-conserving mechanisms is that they are an integral part of the living native forest. As long as the naturally occurring forest is relatively undisturbed, the nutrient-conserving mechanisms remain intact and their recycling function is not interrupted. However, when the forest is cut and burned in preparation for cultivation or pasture, the nutrient conservation mechanisms are destroyed. As a result, the nutrients incorporated in the organic matter of the forest are either volatilized,

Table 3.2 Volume weighted nutrient concentrations (mg/l) in the throughfall of the Oxisol and Spodosol site and world average from Parker (1983).

	Oxisol Site	*Spodosol Site*	*World Average*
Ca	0.10	0.19	2.58
K	0.50	0.83	3.72
Mg	0.04	0.11	1.39

Table 3.3 Caloric values of leaves of selected species from San Carlos (from M. Maass, unpub.).

	$\bar{X}$ *(kcal/g.d.w.)*	*s.d.*
Spodosol Forest:		
Clusia sp.	6.2	0.6
Couma catingae	5.6	0.3
Micrandra sprucei	5.7	0.7
Eperua leucantha	5.7	0.4
site average	5.8	
Oxisol Forest:		
Licania heteromorpha	5.1	0.5
Ocotea costulata	5.4	0.7
Ormosia sp.	5.6	0.2
Protium sp.	5.0	0.7
site average	5.3	

leached into the mineral soil where they can be fixed in unsoluble form (phosphorus), or leached out of the root zone (some cations). The rate at which these losses take place during slash and burn agriculture, and the effect of losses on primary productivity, are the subjects of a following chapter.

However, before a discussion of the impact of cutting and burning on the native forest at San Carlos, it is important to put its structure and function in perspective by a comparison with other tropical forests. Such a comparison gives an indication of the extent of occurance of the nutrient-conserving mechanisms. This in turn could indicate the relative nutrient losses and consequent decrease in productive potential which may be expected following deforestation at each of the sites compared.

Table 3.4 Per cent decrease in concentration of nutrients before leaf fall in three sites at San Carlos ($\pm$ 1 S.D.) (From Sprick, 1979).

Species	Per cent Decrease		
	Phosphorus	*Nitrogen*	*Potassium*
Oxisol			
Protium sp.	53.5	6	71
Licania sp.	74.5	47	59
Aspidosperma megalocarpum	74	22	47
$\bar{x}$	67.3 ± 12.0	25.0 ± 20.7	59.0 ± 12.0
Spodosol – Caatinga			
Manilkara sp.	57	47	73
Eperua leucantha	45	45	75
Micrandra spruceana	72.5	60	78
Eperua purpurea	79	35	88
Guatteria sp.	54	54	74
Mabea sp.	59	51	65
Gustavia sp.	47	31	54
$\bar{x}$	59.1 ± 12.6	46.1 ± 10.3	72.4 ± 10.6
Spodosol – Bana			
*Macairea rufescens**	81	53	67
Aspidosperma album	39.5	23	85
Remigia involucrata	56	34	75
$\bar{x}$	58.8 ± 20.9	36.7 ± 15.2	75.7 ± 9.0

CHAPTER 4

COMPARISON OF FORESTS AT SAN CARLOS WITH THOSE IN OTHER TROPICAL REGIONS

As has already been shown, studies of the characteristics of communities of the forests near San Carlos have suggested that they are under stress, and that the most probable cause of this stress is low nutrient availability. The effect on the animal communities is reflected in the low population density, while on the plant communities it is reflected in the somewhat lower biomass and net primary productivity in comparison with other tropical forests, as well as in the presence of nutrient-conserving mechanisms such as a high root/shoot ratio and relatively thick, long-lived leaves.

The impression gained from these comparisons is that the forests at San Carlos are unique, and that nutrient scarcity is the most important cause of the uniqueness. However, all ecosystems are unique in one way or another. We have not so far developed a picture of how the San Carlos forests fit into the wider spectrum of lowland tropical forests, since there has not been a quantitative evaluation of the range of differences in structure and function of these forests, and the importance of nutrient scarcity in contributing to this range.

One way to compare the forests at San Carlos with other tropical forests is through the use of ordination techniques. In this chapter, ordinations are used to answer question three of the introduction: "How do the forests at San Carlos compare with other tropical forests?".

A. FORESTS COMPARED

Comparisons of nutrient stress and adaptations to it among tropical forests should not include forests that are stressed by different factors – for example, cloud forests stressed by low temperature and high humidity, dry forests stressed by lack of moisture, or mangrove forests stressed by salt water.

Since ecosystem structure and function also depend in part on the successional state, ideal comparisons would include only mature forests, and therefore it follows that comparisons should include only lowland, evergreen, broad-leaved, undisturbed, mature tropical forests with no pronounced dry season. However, finding enough studies which fit these criteria and which also include sufficient parameters for comparisons is difficult, so, in a few cases, criteria were necessarily broadened slightly.

Two of the sites used in the ordinations are the mixed forest on Oxisol and the high caatinga forest on Spodosol at San Carlos. The other five sites are listed in Table 2.4: El Verde, Puerto Rico, at an altitude of 500 meters, is not really a lowland forest and is not really a primary forest, but rather a lower montane late successional forest; the Banco forest on the Ivory Coast has two dry seasons per year, but is still evergreen; the Pasoh forest in Malaysia is characterized by the abundance of very tall trees of the family Diptero-carpaceae; the forest at La Selva, Costa Rica is on very deep soils of alluvial and volcanic origin, and the Panamanian forest is on soils derived from rock high in calcium.

B. ORDINATION METHODS

The techniques used were polar ordination, reciprocal averaging and centred and standardized principal components' analysis (Gauch, 1977; Gauch *et al.*, 1977) and details of the sites ordinated and the parameters used are listed in Tables 2.2 and 2.3. Not all the parameters at every site were measured, but wherever a value is missing, the value used in the ordination is the average over all the other sites where values were available. This procedure minimizes bias, which could be greater were assumed values or zero values used.

The first thirteen variables listed are biotic, (Table 2.2) and the last five are soil (Table 2.3). Ordinations were carried out with and without the soil variables. In Table 2.2, parameter 2 (above-ground biomass) was not included in the ordinations because these data are included in parameter 3 (the root/shoot ratio). Parameter 7 (the predicted leaf biomass) was also not included because it was based on the two previous parameters. Parameters 2 and 7 are therefore retained in this table for discussion purposes.

C. RESULTS AND DISCUSSION

1. Biological Variables

The first axis ranking for both the centered and standardized principal components' analysis and the reciprocal averaging, together with relative scores for the biological variables (Table 2.2) are given in Table 4.1. Both techniques resulted in the same order of ranking with the exception of the

Panama forest, which was last in the principal components' analysis but between the Banco and Pasoh forests in the reciprocal averaging method. Polar ordination resulted in the same rank order as the principal components analysis. For this analysis, the first axis accounted for 64 per cent of the variance among the ecosystems, and for the reciprocal averaging it accounted for 73 per cent. Each subsequent axis for both techniques accounted for 17 per cent or less of the variance, and these axes were probably ecologically meaningless.

The reasons for the rankings become clearer as each biological parameter in Table 2.2 is considered. In order to see the trends more clearly, the sites have been ordered from left to right in the same order as their ranking by the principal components analysis. Root biomass (parameter 1) decreases regularly in the ranking from Caatinga to Panama forest, except for an inversion of the San Carlos Oxisol and El Verde sites. Relatively high root biomass can be taken as an indicator of nutrient-poor sites when flooding or subsoil hardpan are not present (Chapin, 1980). Above-ground biomass (parameter 2) does not show a strong trend along the gradient, but the root/shoot ratio (parameter 3) parallels the root biomass gradient. The presence of roots in a mat above the soil surface (parameter 4) occurred only in the San Carlos ecosystems, although individual roots often occur above the mineral soil in other sites (personal observation in all but the Banco sites). As has been already pointed out, roots close to the surface may confer a selective advantage in obtaining nutrients from decomposing litter in areas where soil nutrient availability is low.

Specific leaf area (parameter 5), the ratio of the leaf area to weight, follows the same trend for Caatinga to Panama, with only San Carlos Oxisol and El Verde inverted. Leaves that have a low area to weight ratio can be indicative of stressful environments (Grubb, 1977). Thick leaves are often

Table 4.1 First axis ranking and scores of ecosystems in Table 2.2 according to centered and standardized principal components analysis (PCA) and reciprocal averaging (RA).

Ecosystem	Score	
	PCA	*RA*
San Carlos, Spodosol	100.0	100.0
San Carlos, Oxisol	71.5	87.5
El Verde	45.8	42.3
Banco	15.7	23.6
Pasoh	13.4	14.0
La Selva	3.2	0.0
Panama	0.0	14.9

called scleromorphic or sclerophytic. Along a fertility gradient in the Brazilian cerrado region, scleromorphic features of the vegetation were positively correlated with low levels of extractable phosphate in the soil (Goodland and Pollard, 1973). In Jamaica (Loveless, 1961, 1962) and in Australia (Beadle 1962, 1966) low soil phosphate also was strongly and positively correlated with scleromorphic features of the vegetation.

Leaf area index (parameter 6) increases regularly from Caatinga to Panama. If specific leaf area and leaf area index are considered simultaneously, another interesting trend emerges: the weight per unit of leaves (reciprocal of specific leaf area) times the leaf area index times an area factor predicts the total leaf biomass of each site. When this is done (parameter 7) no trend is apparent, and the sites do not seem to differ from one another in quantity of leaves. The uniformity of predicted leaf biomass is striking. The reason for the uniformity may be that where leaf index is low, the leaves are relatively thick (low specific leaf area), while where leaf area index is high, the leaves are thin.

Leaf litter production generally increases from San Carlos to Panama, but the increase is not regular (parameter 8). However, when leaf biomass (parameter 7) is divided by leaf production (parameter 8) to obtain the turnover time of leaves, the ratio along the gradient decreases almost regularly (parameter 9). If long-lived leaves are an adaptation to stress, as indicated by Chapin, (1980), the data suggest that the San Carlos forests are most influenced by stress.

There is no clear trend in wood production rates (parameter 10), because data are lacking. The decay constant for leaf litter (parameter 11) is lowest in the two San Carlos forests, indicating that decomposition there is slowest, probably attributable to the same factors that cause the toughness and longevity of the leaves. The value of the decay constant k increases regularly in the principal components analysis ranking, except for a slightly lower value at the Panama site.

In the previous chapter on stress in the San Carlos forest, it was suggested that nutrient concentration in leaf litter, or its inverse, nutrient-use efficiency, could be an indication of the nature of the stress found in some of the communities. Except for the Caatinga and Banco sites, the phosphorus-use efficiencies (biomass to phosphorus ratios, parameter 12) decrease regularly along the gradient, suggesting that phosphorus may play a role in establishing the gradient shown by ordination. In contrast, the Caatinga site has a high biomass/nitrogen ratio, suggesting that, there, nitrogen is the limiting element.

2. Soil nutrient variables

The examination of nutrient-use efficiency in leaves in Table 2.2 suggested that nutrients are an important variable in establishing the ranking. However, when soil nutrient factors (Table 2.3) are included in the ordinations, the variance which could be accounted for by the first axis dropped to 50 per cent for both ordinations. The problem may be that the nutrients analyzed may not be the critical nutrients, or the methods used may not have been appropriate.

The first two parameters in Table 2.3 are exchangeable calcium and total calcium in the soil. Although the graph in Fig. 2.6 suggested that calcium may be important in the Oxisol site at San Carlos, it also suggested that phosphorus was a more common limiting nutrient at other forest sites in the tropics.

Phosphorus is another of the parameters in Table 2.3, but the values are for total phosphorus, which bears little relationship to the amount of phosphorus that plants can take up – that is, the available phosphorus. Soil pH in Table 2.3 also bears little relation to the rank order. In contrast, total soil nitrogen does follow the same relative rank order, where data are available, but it does not correlate well with the absolute scores (Table 4.1). Like phosphorus, the rate at which nitrogen becomes available to plants is probably more important than the total amounts in the soil. Rates of decomposition are probably a better indicator of nitrogen availability than total nitrogen, and decomposition constants, which indicate rates at which nitrogen is mineralized (row 11, Table 2.2), correlate well with absolute scores.

Soil water stress also may be a factor which occurs in the Caatinga site but not in the other sites. The deep coarse sands of the Caatinga do not hold moisture effectively in the root zone and water stress can occur within a few days after a rainstorm (Sobrado and Medina, 1980).

3. Ecosystem classification

The examination of ecosystem characteristics along the ranking indicated by the ordination suggests that no single parameter alone is responsible for the environmental gradient represented by the first axis of ordination. Available phosphorus seems a likely factor, but the rate of nitrogen release during decomposition is also probably important. In addition, water may play a role, at least in the Caatinga forest.

It would be convenient to have simple terms to distinguish forests characterized by highly developed adaptations to nutrient stress from forests which show little reaction or adaption. Whitmore, (1975) has used the term 'oligotrophic' to refer to the heath-like forests of Southeast Asia, called Kerangas, and which are structurally similar to the Caatinga forest. Both forests have scleromorphic leaves, low stature and thick root mat, as well as the low soil nutrient levels. Both usually occur on podzol soils. Application of the term oligotrophic to forests can be criticized, because both that term and its opposite, 'eutrophic', have limnological connotations. Nevertheless, rather than coin new terms, it might be preferable to refer to ecosystems at the Caatinga end of the gradient as oligotrophic and to those at the Panama-La Selva end as eutrophic.

Practical implications of a fertility classification

The ordination of ecosystems from oligotrophic to eutrophic has important practical implications. It gives insights into land use capability that is not evident from other types of plant community classification. For example, Holdridge, (1967) classified plant formations on the basis of climatic parameters. Thus, using this method, Amazonian and Central American forests are both classified as tropical moist or tropical wet forests. In classifications based on biomass, both are tropical broad-leaved humid forests (Olson *et al.*, 1982). There is nothing in these classification schemes to suggest any difference in the agricultural potential of the two regions, and even casual observations by a visitor to both forest types may not suggest important differences. For example, the similarity of the biomass of forests in the Amazon and lowland Central America suggests to the inexperienced observer that the two regions may not differ significantly in their ability to sustain agricultural production when the forests are cut and converted to agriculture. However, it is well known by tropical agronomists and ecologists that the two regions differ immensely in natural soil fertility. Because most of Central America is much younger geologically than the Amazon region, soils are less weathered and leached, and nutrient levels in many Central American soils are relatively fertile compared to soils of the Amazon. These differences are reflected in characteristics of the native forests: for example, the amount and distribution of roots and the degree of sclerophylly of the leaves.

E. CONCLUSION

This chapter compared the forests at San Carlos with other tropical forests through the use of ordination techniques. The ordinations suggested that the forests at San Carlos represented one end of the spectrum of tropical forests. The forest growing on Spodosol showed the highest development of characteristics of what can be called an oligotrophic ecosystem – indeed, it might even be called the prototype oligotrophic forest ecosystem – and the forest growing on Oxisol was second in the development of oligotrophic characteristics.

Those soils falling into classification orders such as Spodosols and Oxisols are not always completely uniform in their fertility – that is, forests growing on Oxisols in other parts of the Amazon may not be as highly stressed as the one at San Carlos. Nevertheless, Oxisols generally are nutrient-deficient compared with soils derived from volcanic activity, alluvium or recent geologic uplift. Most tropical forests growing on Oxisols, be they in South America, Southeast Asia or Australia, might be expected to be on the oligotrophic end of the spectrum and have the symptoms and adaptations of nutrient-stressed forests.

CHAPTER 5

PRODUCTIVITY AND NUTRIENT DYNAMICS DURING SLASH-AND-BURN AGRICULTURE

How do the nutrient cycles and net primary productivity of a tropical forest ecosystem change during slash-and-burn agriculture? This is the question now to be addressed in this chapter. The results of the ordination in the last chapter suggest that the impact of cutting and burning may vary greatly depending on the type of ecosystem. In eutrophic ecosystems, such as those the relatively rich soils of Central America, forests lack highly developed nutrient conserving mechanisms, and therefore cutting and burning might not have a readily discernible impact either on nutrient retention, because of the high exchange capacity of the soils, or on primary productivity, because of the high nutrient stocks of the soil.

In contrast, the forests at San Carlos show highly developed nutrient conservation mechanisms and, as already suggested at the end of the last chapter, the forest growing on Spodosol might be characterized as the prototype oligotrophic forest ecosystem, while the forest growing on Oxisol was second in the ordination in the development of oligotrophic characteristics. Because cutting and burning of these forests destroys the nutrient-conserving mechanisms, the practice might be expected to result in relatively high nutrient losses and consequent decreases in productivity.

Because high nutrient losses seem more probable following the cutting of an oligotrophic forest than a eutrophic forest, it was felt that an experimental slash-and-burn cultivation in an oligotrophic site might better resolve the puzzle posed by the studies cited in the Introduction: namely, that, despite the general idea that cutting and burning of tropical forests results in nutrient losses, there is little evidence that slash-and-burn agriculture results in important nutrient losses since most of the studies reported higher levels of nutrients in the soil at the time cultivation was abandoned than in the soil of the pre-cut forest.

A. METHODS

The field site chosen was located four kilometers east of the village of San Carlos and an experimental plot and control plot, each one hectare in area, were established to determine the changes in nutrient cycles and productivity during slash-and-burn agriculture. The plots were on an Oxisol, because this is the soil type often used by local cultivators. (Spodosols are almost never used.) The two plots were on the same Oxisol hill, and were about 50 meters apart. Fig. 5.1 is an aerial view of a similar slash-and-burn site on an Oxisol hill near the study site.

Fig. 5.1. Aerial view of the terrain near San Carlos showing a recently abandoned agricultural site near the river and a recently cut and burned site near the center of the photograph. Both sites are on an Oxisol hill. A lower area, supporting caatinga forest on Spodosol, runs to the left of and below the recently burned site.

The control plot was used to determine if any variations in productivity or nutrient dynamics during cutting and cultivation in the experimental plot were attributable to variables external to the experiment. The forest biomass and the standing stock of nutrients were determined for the control plot in 1975, the nutrient input–output balance was measured from September, 1975, to July, 1983, and the net primary productivity of forest trees was measured between July, 1975, and July, 1983.

The biomass and the standing stock of nutrients on the experimental plot were determined before cutting, and at intervals of a year or less thereafter. The nutrient input–output balance was measured simultaneously with the control plot. The cutting of the experimental forest occurred in August and September of 1976 (Fig. 5.2), and the experimental plot was burned in December of that year (Fig. 5.3) and cultivated from January, 1977, to January, 1980. The net primary productivity of crop and 'weed' (successional)

Fig. 5.2. Cutting the experimental site in September, 1976.

71

Fig. 5.3. Burning of the experimental site in December, 1976.

species was measured in the experimental site from the start of cultivation, through abandonment in 1980, until 1983. Crop productivity was separated into edible and total productivity. Details of the methods used to determine biomass, productivity, and nutrient balance in both the control and experimental sites are given in Appendix C.

Immediately after the burn, half of the one hectare experimental plot was protected from further disturbance. This plot was used for the successional studies reported by Uhl *et al.* (1981) and Uhl and Jordan (1984). The other half of the experimental site was planted with local crops and, of that half-hectare, one plot of 30 × 50 meters was designated for the intensive study of productivity and nutrient balance.

Forest cutting and burning, together with planting, weeding and harvesting of the slash-and-burn site were carried out by local workers under the supervision of an experienced local farmer to ensure that operations conformed with local practice. The main crop planted was yuca, or manioc (*Manihot esculenta* Crantz), interspersed throughout with pineapple (*Ananas comosus* L. Merr.), plantain (*Musa* sp.) and cashew (*Anacardium occidentale* L.).

Yuca is a common staple root crop throughout the tropics, and is an important crop for indigenous cultivators in the Amazon region. In San Carlos, the 'sour' variety is preferred, because of its resistance to herbivory and decomposition. The root can be left in the ground until it is convenient to harvest. The mechanism of protection is hydrocyanic acid generated from cyanogenic glucosides when the plant cells are ruptured (Montagnini and Jordan, 1983). These poisonous compounds are extracted during the course of the preparation of flour (Uhl and Murphy, 1981).

The first harvest of yuca began late in 1977. Planting and harvesting in a slash and burn farm plot, or 'conuco' as they are called locally, are almost continuous processes – the planting of the second crop begins before the first is fully harvested. During its lifespan, there were three complete crops of yuca harvested from the experimental plot, pineapples were harvested whenever they were ripe, plantain never bore fruit, and cashew began to bear fruit only during the third year.

B. RESULTS

1. Productivity

The productivity of the control forest varied only slightly from year to year (Table 5.1), suggesting that external variables, such as climate, may have had little influence on changes observed in the slash-and-burn plot throughout the experiment.

In the experimental conuco, total productivity actually increased during the second year of cultivation, but during the third it dropped well below the level of the first year. When the productivity of yuca alone is considered, this dropped gradually over the three year period, but the production of edible roots fell by about fifty per cent (Table 5.1). Photo of the conuco, just before the beginning of the first harvest (Fig. 5.4) and during the third cropping cycle (Fig. 5.5), illustrate the decline of the above-ground standing crop.

2. Nutrient dynamics

a. Leaching losses

The concentrations of nutrients in the leachate during and after cultivation are compared with the concentrations in the leachate from the control plot in Fig. 5.6. Concentrations from the area of the experimental plot that was burned and immediately abandoned to successional vegetation are included in the figure for comparison. The results of tests showed that concentrations of calcium, potassium, magnesium and nitrate nitrogen in the leachate water of the experimental plot increased after the cut and then increased sharply

73

Table 5.1 Net primary productivity of control forest during the one year pre-cultivation period and three years of cultivation, and of the experimental site during cultivation.

Year	Pre-cult.	*kg/ha/yr, dry wt* 1	2	3
Control Site				
Wood	3622	4724	5037	5255
Leaves	6344	6008	5948	5655
Roots*	2010	2010	2010	2010
Total control	11967	12742	12995	12920
Experimental Site				
Yuca, stems, leaves, fine roots		3485	3278	1838
Yuca, edible roots		1465	1006	700
Yuca, total		4950	4284	2538
Other crops, total		383	1010	612
"Weeds", total		300	679	990
Total, experimental		5633	5973	4140

* Root productivity determined only once.

Fig. 5.4. The experimental plot in September, 1977. The major crop is yuca. Several individuals of Cecropia are in the foreground.

Fig. 5.5. The experimental plot in January, 1979.

after the burn. They gradually decreased during the period of cultivation, but there were periodic increases which were the result in most cases of low volumes of drainage water, usually occurring during the drier months from January to March. By the time cultivation was abandoned in January, 1980, concentrations were close to the levels in the control forest.

It is not clear what caused the increase in ammonium nitrogen concentration just after cutting and burning. The concentrations in the conuco immediately after the cut suggest a response to it, but concentrations in the forest control plot showed an increase between 1977 and 1978 that was almost as large as those in the experimental plots.

In contrast to the other nutrients, no phosphorus leaching losses could be detected, either as phosphate or organic phosphorus. Phosphorus dynamics between soluble and bound forms in the soil appear to have been much more important, and these are discussed later.

The rates of nutrient leaching (Fig. 5.7) are the product of the concentrations of nutrients in Fig. 5.6 and the rate of water percolation out of the root zone, determined by the water budget described in Appendix C. The change in leaching rates in the experimental plot after cutting and burning is more pronounced than for concentration alone, especially for calcium. The reason is that, compared to the control plot, both concentrations of nutrients in the water and the volumes of leachate are higher in the experimental plot, the latter because the crop transpires much less than does the undisturbed forest and, as a result, much more water percolates through the soil of the crop ecosystem.

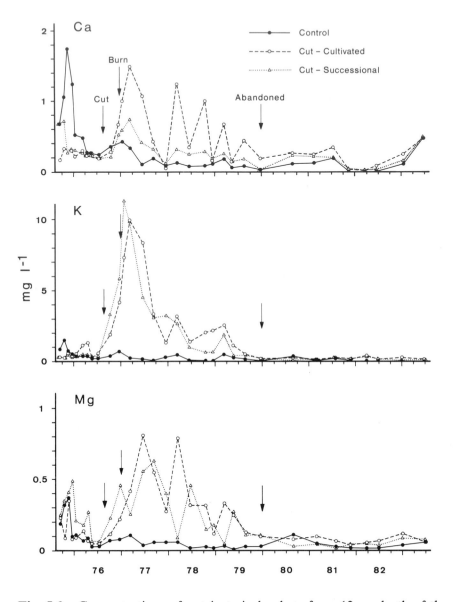

Fig. 5.6. Concentrations of nutrients in leachate from 12 cm depth of the control and cultivated plots and from the plot that was cut, burned, and abandoned.

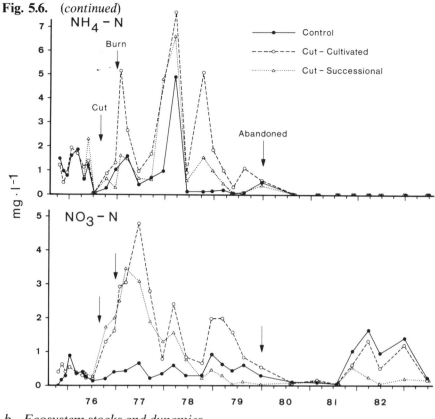

Fig. 5.6. *(continued)*

b. Ecosystem stocks and dynamics

The stocks and dynamics of nutrients in the experimental site are illustrated in Figs. 5.8 and 5.9. The vertical axis above zero shows the ecosystem standing stocks of each nutrient. The compartments are stacked so that the total ecosystem stock is the sum of the stocks in each of the compartments. The stocks to the left of the cut and burn arrow are the quantity of nutrients in each compartment of the forest before cutting, taken from Table 5.2. Changes in stocks after cutting and burning and during cultivation are shown to the right of the cut and burn arrow.

Cumulative net losses from the ecosystem are plotted below zero on the vertical axis in Figs. 5.8 and 5.9. Since the nutrient input-output balance of the control forest was close to zero (Fig. 2.1) net losses were therefore assumed to equal zero in the experimental plot before cutting. Cumulative losses were calculated starting at the time of the cut. Total cumulative losses at the end of each year are the sum of the total losses for the previous years plus net losses or gains for the most recent year.

There are three types of net nutrient losses following cutting and burning. One is the excess of nutrients leached from the soil over nutrients moving into the ecosystem through bulk precipitation. A second is the excess of

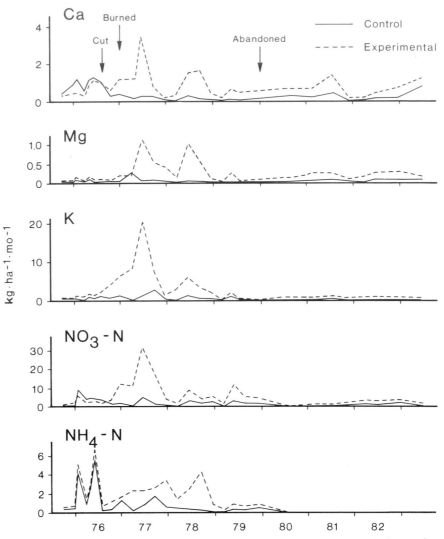

Fig. 5.7. Rates of nutrient leaching in the control and cultivated plots. The rates are normalized to a monthly basis.

denitrification over nitrogen fixation. The third is the excess of biomass stocks removed from the plot over stocks imported. Stocks removed from the plot are primarily harvested edible crops. Importation is stem stocks for planting.

The increase in soil stocks of calcium, magnesium, and to a lesser extent potassium following the burn (Figs. 5.8 and 5.9) illustrates the conversion of these nutrients by fire from a bound form in the tree biomass to a soluble form in the soil. The relatively small but abrupt drop in total ecosystem nitrogen directly below the arrow signifying the cut and burn represents an unaccounted for loss, probably due to volatilization.

There are four major trends illustrated in Figs. 5.8 and 5.9 that will be important to the discussion of productivity and nutrients during cultivation:

1. Soil nutrient stocks increase after burning, and levels remain higher than in the pre-cut forest throughout cultivation, except for a dip by nitrogen during the second year which was reversed during the third.

2. Throughout cultivation, the stocks of nutrients in the crops and successional vegetation are low compared to the total amount in the ecosystem.

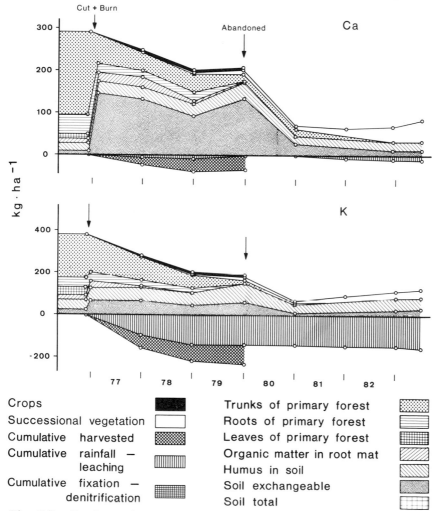

Fig. 5.8. Stocks and cumulative losses of calcium and potassium as a function of time in the experimental plot.

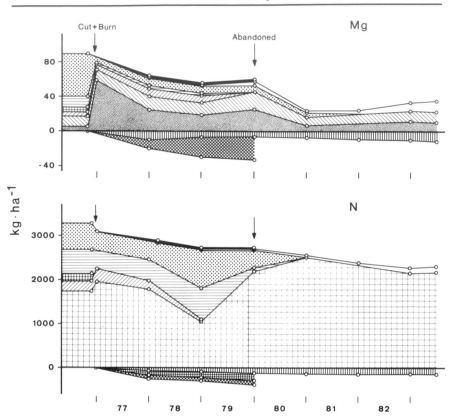

Fig. 5.9. Stocks and cumulative losses of magnesium and nitrogen as a function of time in the experimental plot. Key as for Fig. 5.8.

3. During the cultivation period, losses from the entire ecosystem are greater than from the soil compartment alone. This is because nutrients lost from the soil through leaching were replaced by nutrients leached down into the soil from the decomposing remains of the primary forest.

4. At the time of abandonment of cultivation, the remains of the original forest biomass, including the organic matter in the root mat, were largely gone.

Dynamics of phosphorus were considerably different from dynamics of the other nutrients. The proportion of total ecosystem phosphorus in the soil of the pre-burn forest (83.7 per cent, Table 5.2) was considerably higher than that of other nutrients. No statistically significant (90 per cent level) changes in total soil phosphorus could be detected throughout the experiment. Only during the latter part of the experiment was it realized that the most important phosphorus dynamic was not leaching loss, nor between soil and

Table 5.2 Total stocks of nutrients in the control and experimental sites.

kg/ha

	Calcium		Potassium		Magnesium		Phosphorus		Nitrogen	
	Control	Experimental	Control	Experimental	Control	Experimental	Control	Experimental	Control	Experimental
Leaves (trees and lianas)	12.0	12.8	39.2	41.9	6.4	6.8	5.3	5.7	166.1	177.2
Wood*	184.7	193.3	197.1	206.5	47.0	49.2	24.8	26.0	746.9	782.0
Roots	52.4	46.0	49.6	43.6	13.4	11.8	19.5	17.1	619.9	545.4
Total Living	249.1	252.1	285.9	292.0	66.8	67.8	49.6	48.8	1532.9	1504.6
Fallen Trunks	4.9	4.9	2.5	2.5	1.9	1.9	0.8	0.8	94.8	94.8
Humus in mat	5.0	3.5	12.2	8.5	2.8	2.0	3.0	2.1	159.0	110.8
Soil, exchangeable	6.9	7.1	22.7	23.3	5.2	5.3				
Soil, organic matter	20.9	21.4	50.7	51.9	11.6	11.9				
Soil, total digest,							242.5	248.6	1697.0	1740.0
Total soil, including humus	37.7	36.9	88.1	86.2	21.5	21.1	246.3	251.5	1950.8	1945.6
Total ecosystem (living plus soil)	286.8	289.9	374.0	378.2	88.3	88.9	295.9	300.3	3483.7	3450.2
Soil/ecosystem (%)	13.1	12.8	23.6	22.8	24.3	23.7	83.2	83.7	56.0	56.4

● Weighted bark, heartwood, and sapwood of stems and branches plus twigs plus lianas.

vegetation, but within the soil between relatively soluble and insoluble forms of phosphorus.

To obtain data on the dynamics of phosphorus in the soil during cultivation, samples were taken in 1984 from the control plot, and a series of cultivated plots of varying age since cutting and burning. All plots were on Oxisol sites near San Carlos.

Phosphorus that is extractable (soluble) in an acid solution is often considered an approximation of the phosphorus that is readily available for rapid plant uptake (Sanchez, 1976). Extractable phosphorus ranges from about two per cent of total phosphorus in the control and abandoned site, to other five per cent in the cultivated site (Table 5.3). Increase in the extractable phosphorus is paralleled by an increase in soil pH, as is usually the case in acid tropical soils (Sanchez, 1976). Soil pH values from the slash-and-burn site during the experiment (Table 5.4) are probably an index of the solubility of soil phosphorus during the experiment.

Table 5.3 Phosphorus concentrations and pH of soil from four sites on Oxisol, Jan. 1984.

	Control	One month since burn	20 months since burn	Four years after abandonment (7 years since burn)
Acid extractable P, ppm	4.80 ± 0.56	7.40 ± 1.52	13.00 ± 11.80	4.70 ± 0.59
Soil pH, water	4.25 ± 0.10	4.96 ± 0.19	5.15 ± 0.55	4.52 ± 0.08
Soil pH, KCl	3.80 ± 0.12	4.19 ± 0.10	4.36 ± 0.50	4.12 ± 0.08
Total P, mg/g	0.20 ± 0.03	0.13 ± 0.01	0.25 ± 0.04	0.30 ± 0.02
Per cent available P	2.4	5.7	5.2	1.6

C. DISCUSSION

1. Productivity

The productivity of crop plants, especially the edible portions, declined, as is almost universally found in slash-and-burn agriculture (Norman, 1979). Most published values for yuca production are in terms of fresh weight of edible roots. The fresh weight values for the three years of cultivation during the San Carlos experiment were 4.31, 2.81, and 2.00 tons per hectare respectively. These values are considerably lower than wet weight yields from other regions. For instance, in Ghana, annual yields ranged from 11-34 tons per hectare (Doku, 1969). The average yield in some regions of

Table 5.4   Soil pH in the experimental site between 1975 and 1983.

Date	pH
Undisturbed forest 4/76	3.87 ± 0.22
Post burn 6/77	5.41 ± 0.29
10/77	4.67 ± 0.30
4/78	4.67 ± 0.11
10/78	3.93 ± 0.17
7/79	4.08 ± 0.38
6/80	4.09 ± 0.53
9/82	4.13 ± 0.07
6/83	3.81 ± 0.12

Brazil were greater than 12 tons (Normanha, 1970), and in northern Venezuela it averaged about 7 tons per year (Petriceks, 1968). The world average production of yuca roots has been estimated at 8.4 tons per hectare annually (Kay, 1973). These lower rates of agricultural production at San Carlos may be a reflection of low soil fertility.

Although the production of edible roots declined steadily throughout cultivation of the experimental plot at San Carlos, the total net primary productivity increased the second year, before decreasing the third, and weed production increased steadily through the three years.

2. Nutrient stocks and dynamics

The increase in nutrient concentrations in the soil following cutting and burning was expected. What was surprising was that the results did not confirm the hypothesis stated by Richards in the introduction – that clearing the forest and cultivating the land would result in rapid losses of nutrients. Instead, stocks of soil nutrients remained high throughout cultivation.

An examination of other studies of slash and burn agriculture in various regions, however, also shows that nutrient stocks generally remain high throughout the period of cultivation. For example, in Ghana on a 'reddish yellow latosol', Nye and Greenland, (1964) found that exchangeable nutrient cations and Kjeldahl nitrogen increased after burning and then decreased but, after two years, levels were still not below those in the soil before forest clearing. Zinke *et al.*, (1978) studied nutrient content in 'latosols' in Thailand at various stages of a ten-year rice–tobacco–cotton rotation system. They found that exchangeable calcium and magnesium were highest in recently cut and burned plots, and that levels of these nutrients remained higher throughout the ten-year cycle than in a nearby 'old-growth' forest that served as a control.

Studies in the neotropics have shown similar trends. Denevan, (1971), working in slash-and-burn sites in eastern Peru, found that nutrient levels in the soils increased following forest clearing and then declined slightly. He questioned the importance of nutrient loss in the decline of crop productivity, stating: "In general, though nutrient levels declined somewhat, with probably corresponding decreases in crop yields, there is no evidence that this was the main reason for the abandonment of fields after only one or two years". Harris, (1971) examined a series of varyingly aged slash-and-burn sites in the upper Orinoco region of Venezuela. His comparison of the mean nutrient values of forest soil samples with the mean values for slash-and-burn soil samples showed that organic carbon, available phosphorus, and exchangeable calcium and magnesium all were higher in cultivated plots. Only exchangeable potassium and sodium were lower than in forest soils. Brinkmann and Nascimento, (1973) studied changes in 'latosols' in the central Amazon Basin for the first year of slash-and-burn agriculture. They found that exchangeable calcium and magnesium in farm plot soils were substantially higher than in soils of the undisturbed forest, potassium initially was higher and then decreased to about the level of the forest and total phosphorus was generally unchanged, except in one flooded plot. Finally, on Ultisols in the Amazon Basin of Peru, in a plot cultivated but not fertilized, it took 5-8 years for most nutrients except nitrogen to decline to the pre-burn level (Sanchez et al., 1983). The relatively high quantities of total nutrients in the soil at the time of abandonment of cultivation at San Carlos then, are not apparently unusual.

The second somewhat surprising result was the relatively large stocks of nutrients in the soil throughout cultivation and at the time of abandonment compared with the stocks in the crops and successional vegetation. A simple comparison of total stocks in the soil and crops would not suggest that lack of nutrients was the cause of declining crop production. In fact, the total amounts in the soil are several times those necessary for another crop, which suggests that the critical factor is probably the proportion of nutrients, especially phosphorus, which are available for plant uptake.

a. Soil pH

One of the most common fertility problems on the infertile, iron- and aluminum-rich Oxisols and Ultisols is phosphate adsorption (Fox and Searle, 1978). Soil acidity plays an important role in this process. During weathering, nutrient cations tend to be replaced by hydrogen or aluminum, with a resultant decrease in soil pH (Uehara and Gillman, 1981). Soil pH governs phosphate reactions with iron, aluminum and manganese and, as soil pH drops, these metals react with soluble phosphorus to form insoluble hydroxy phosphates. In silicate minerals such as kaolinite, hydrous oxides of iron, aluminum and manganese occur on the clay surfaces, where they readily react with phosphate to form hydroxy phosphates (Brady, 1974). Because many soils in the humid tropics are highly weathered, acid, relatively high in iron and aluminum and contain a large proportion of silicated minerals, phosphorus fixation is often an important phenomenon.

As a result of burning the slash during shifting cultivation, the ash, rich in basic cations, enters the soil and raises soil pH. The effect appears to be the same as lime application, a practice common on acid agricultural soils of temperate regions, and the availability of phosphates for crop uptake increases.

A problem at the San Carlos experiment with the idea that bases regulate soil pH, which in turn regulates phosphorus availability, is that throughout the cultivation period soil pH and available phosphorus declined, but input of calcium, potassium and magnesium from decomposing slash into the soil exceeded leaching losses from the soil (Table 5.5). Although there were year-to-year variations, over the 1977-1979 cultivation period beginning after the burn and ending at abandonment, the sum of the three-year totals indicates that there should have been a net gain of these nutrients in the soil, if harvesting losses are not important. In the case of magnesium, harvest losses are important but, nevertheless, the stock of magnesium remained well above the level of the forest before cutting.

b. Organic matter and phosphorus solubility

A different factor which may have been important in regulating plant productivity is decomposing organic matter, or the organic materials leached from decomposing organic matter. The decline in productivity of slash-and-burn agriculture on the Oxisol paralleled the disappearance of organic matter on the soil surface due to decomposition of slash during the period of cultivation. Figs. 5.8 and 5.9 give a picture of the rates at which the organic matter disappeared at San Carlos. The quantity of slash remaining in the experimental plot during cultivation is proportional to the declining amounts of nutrients in the remains of the trunks, roots and leaves of primary forest, and organic matter in root mat shown in Figs. 5.8 and 5.9. While these figures indicate that some surface organic matter still remained at the time of abandonment, that which was left was almost entirely trunks. There were large patches of exposed mineral soil between the trunks, due to the complete disappearance of the organic matter in the root mat.

The correlation between decreasing organic matter, both on top of and in the soil, and crop productivity suggests that the organic matter may play a

Table 5.5 Nutrient movement into the mineral soil from the layer of humus and roots (column headed "in"), and leaching out of the mineral soil (column headed "out"). Values are kg·ha^{-1}.

	Calcium				Potassium				Magnesium			
	Experimental		Control		Experimental		Control		Experimental		Control	
	In	Out	In	Out	In	Out	In	Out	In	Out	In	Out
1976	5.89	9.64	5.48	8.04	4.51	11.50	8.97	7.95	2.73	1.44	1.76	1.29
Cut												
1977	7.91	13.16	3.40	2.40	162.55	131.08	9.69	5.01	6.40	9.20	1.71	0.77
1978	18.35	12.65	2.32	1.69	37.57	54.08	4.42	2.34	7.89	6.60	.86	0.38
1979	10.15	5.04	1.82	1.14	29.26	11.28	7.65	2.98	5.15	2.88	.80	0.31
Σ 77-79	36.41	30.85	7.54	5.23	230.08	196.44	21.76	10.33	19.44	18.68	3.37	1.46

	NO$_3$ - Nitrogen				NH$_4$ - Nitrogen			
	Experimental		Control		Experimental		Control	
	In	Out	In	Out	In	Out	In	Out
1976	4.32	15.17	2.23	6.35	32.77	25.18	22.49	16.20
Cut								
1977	9.13	43.97	4.39	5.84	18.25	28.26	17.16	8.71
1978	5.22	14.08	3.08	5.25	14.70	30.79	14.36	1.53
Σ 77-78	14.35	58.05	7.47	11.09	32.95	59.05	31.52	10.24

role rendering phosphorus more available for uptake by crops. There are several ways in which decomposing organic matter could contribute to an increased supply of phosphorus. One is directly through the mineralization of phosphorus in the decomposing organic matter. This process is carried out largely by extracellular enzymes produced by phosphorus-requiring organisms (McGill and Cole, 1981). Phosphorus released from fresh organic matter may enter either a stable organic pool or be released in labile form into the soil (Jones *et al.*, 1984). Some of the labile phosphorus could be adsorbed by organic and inorganic constituents of the soil, but some of it could be available for uptake by plants.

Another way in which the decomposing organic matter could influence the balance between soluble and iron- and aluminum-bound phosphate in soil is through the liberation of phosphate-dissolving compounds. Several laboratory studies (Swenson *et al.*, 1949; Struthers and Sieling, 1950; Dalton *et al.*, 1952) suggested that organic acids liberated during the decomposition of organic matter were responsible for solubilizing phosphate complexed by iron and aluminum. Powell *et al.*, (1980) and Cline *et al.*, (1982) showed that naturally occurring organic compounds in soil were important in chelating iron, and Graustein *et al.*, (1977) showed that calcium oxalate which occurs in the litter of several different soils keeps phosphorus available to plant roots through the chelation of iron and aluminum.

A third way that decomposing organic matter could result in higher levels of available phosphorus is that nutrients and carbon leached from decomposing organic matter stimulate microbial growth and microbial activity, and products to plant roots through the chelation of iron and aluminum.

A third way that deccomposing organic matter could result in higher levels of available phosphorus is that nutrients and carbon leached from decomposing organic matter stimulate microbial growth and microbial activity, or products of microbial metabolism play a role in solubilizing phosphate. Following an early report of this phenomena (Gerretsen, 1948) many researchers examined the relationships between production of organic acids in the rhizosphere and plant uptake of phosphorus from mineral sources, but the mechanisms of solubilization are still not clear (McGill and Christie, 1983).

c. Interaction of phosphorus and nitrogen

The fixation of phosphorus by iron and aluminum at low pH may also have important effects on the nitrogen cycle. Walker and Adams, (1958, 1959) hypothesized that biological nitrogen fixation would cease under natural conditions when available inorganic phosphorus had disappeared and non nitrogen-fixers were competing successfully for all the nitrogen and phosphorus being mineralized from soil organic matter. An examination of a chrono-sequence of soils on alluvial terraces and glacial till deposits produced some evidence for this. Nitrogen fixation appeared to be partially inhibited as phosphorus became less available (Stevens and Walker, 1970). A common practice for increasing nitrogen fertility in the grasslands of Australia and New Zealand is the application of phosphorus fertilizer, which in turn promotes establishment and growth of nitrogen-fixing legumes (Cole and Heil 1981).

Other nitrogen transformation processes such as mineralization, nitrification and denitrification are also strongly influenced by available phosphorus, and rate processes may be limited by phosphorus when phosphorus in a soluble form is not readily available to microbes which mediate these processes (Cole and Heil, 1981).

Nitrogen could affect phosphorus availability, as well as the reverse, and can stimulate the growth of microorganisms which affect the availability of phosphorus in decomposing litter (Stevenson, 1964). It also affects the growth of higher plants, which in turn can affect the availability of phosphorus. For example, phosphorus mobilization from recalcitrant organic matter in the soil is facilitated by oxalic acid produced by mycorrhizae associated with the roots of higher plants (Sollins *et al.*, 1981).

d. Field observations near San Carlos

The literature review suggests an important role of soil organic matter in the solubilization of phosphorus complexed by iron and aluminum, although the nature of that interaction it is not yet clear. It also appears possible that increased phosphorus solubilization results in greater nitrogen availability, and the higher levels of nitrogen and phosphorus are important in stimulating the growth of crop plants.

Informal observations outside the experimental plot suggest that native cultivators already are aware of the benefits of organic matter in stimulating crop growth. It was noticed that these farmers would rake organic litter and slash around the base of the yuca plants when production began to decline. To investigate the significance of this, C. Uhl, (unpublished) established two plots within the experimental conuco, but outside the intensely studied plot, near the end of the three year cultivation period. In one plot, organic matter was raked around the base of newly-planted individual yuca plants, but the other plot was left with the mineral soil surface almost completely exposed. The growth rate of the plants in the plot with surface organic matter produced biomass at a significantly (99 per cent confidence) higher rate than those without.

It is unlikely that the stimulation of growth observed by Uhl resulted directly from a pH effect. In contrast to the situation when the slash is burned, the pH in this experimental plot probably remained low, since there was no burn. The addition of decomposing slash and litter should not have raised the soil pH.

It is also unlikely that the stimulation of growth resulted from mineralization of phosphorus bound in that applied litter. The growth of the plants in the plots with slash was almost as great as that in the plot where the forest was cut and burned, despite vastly different rates of phosphorus release.

It is more likely that the increase in growth resulted either from a solubilizing effect on phosphorus of organic compounds released from decomposing organic matter, or from stimulation of microbial growth due to organic matter leachate. In the second case, products of microbial metabolism might solubilize phosphate. There clearly remain important questions about phosphorus mobilization by organic matter in the nutrient-poor Oxisols and Ultisols of the tropics.

3. Other possible causes of crop decline

In addition to nutrient availability, other possible causes of the crop decline in the San Carlos experiment were also investigated. Climate was ruled out, because in the control forest the yearly increments of above-ground dry weight did not change greatly during the time of the experiment (Table 5.1).

In some slash-and-burn sites, decline in crop production has been attributed to vigorous invasion by weeds (Watters, 1971). In the experimental site at San Carlos, the loss of nutrients to weeds probably was not important, since the experimental plot was weeded thoroughly. Competition for light by weeds can also reduce crop production in Amazonian sites. If no weeding occurs, crop plants are quickly shaded out. However, the crop plants in the experimental concuco were never overtopped by weed vegetation during the period of cultivation and it is therefore unlikely that weeds were responsible for the decline in crop productivity.

Another cause of decreased productivity in some tropical regions is soil erosion, which lowers productivity by removing nutrients and topsoil and exposing denser and less fertile subsoil. Bulk density at the time of abandonment was 1.23 ± 0.15 (1 s.d.) g/cm3, a value not significantly different from the value before forest cutting which was $1.17 + 0.10$ g/cm3 (Appendix C.1), suggesting that changes in bulk density probably were not important in the decline of crop productivity. Attempts at measuring erosion by driving marked pegs into the ground were generally unsuccessful, because soil moved away from one spot seemed to be compensated for by other soil moving into that spot. Our general impression was that as long as there was a continuous cover of decomposing root mat on the soil surface, erosion was not serious. However, after the disappearance of the root mat towards the end of the third year, the importance of erosion may have increased. This will be discussed further in the next chapter.

Insect consumption can also contribute to crop yield declines. At the experimental site, chewing and boring insects consumed only between 2 to 3 per cent of the net primary production (Montagnini and Jordan, 1983). However, up to 14 per cent of the plant's elemental uptake was lost to sucking insects. Price (1975) suggests that sucking insects may achieve a spatial avoidance of plant chemical defenses, as their fine mouth parts enable them to feed between pockets or ducts of toxin in the host plant. Alternatively, sucking insects may be ingesting the cyanogenic compounds in the leaf phloem sap, but do not break the cells and release the enzyme which hydrolizes the glucosides into hydrocyanic acid, hence the toxin does not affect them. The nutrients ingested by the sucking insects probably were not lost from the slash-and-burn plot, but rather returned to the soil, since the species of sucking insects involved appeared to be restricted to cultivated sites (Montagnini and Jordan, 1983).

In some tropical areas, nematodes can influence crop productivity. For example, in the Ivory Coast (Luc, 1968) and in Togo (Guiran, 1965), nematodes frequently affect yuca. However, nematodes seem to be a problem more in areas that are under continuous cultivation. There is little information which would implicate them as serious pests in shifting cultivation sites of the New World tropics. Neither Lozano *et al's* (1976) book *Field Problems of Cassava* (i.e. yuca) nor recent issues of *Nematropica* and the *Cassava Newsletter* suggest nematodes are important in affecting yields of yuca in slash-and-burn agriculture.

Yuca in the experimental conuco also showed no signs of bacterial, viral, or fungal infections which cause diseases such as frog skin disease, African mosiac, superelongation and bacterial stem rot.

The decline in soil pH during cultivation increases the solubility of aluminum in the soil which results in increased aluminum uptake by crop plants. Aluminum toxicity is often a problem for crops growing in aluminum rich Oxisols and Ultisols (Sanchez, 1976).

Although it is not possible to state the reason for the decline in crop production at the experimental site with absolute certainty, the most important factor may be the declining levels of available phosphorus. Phosphorus decline seems not to be due to loss from the ecosystem, but rather to a change in form from relatively soluble to insoluble. This change may be due to a decrease in soil pH, and/or to a decrease in organic compounds that are leached from soil organic matter and that are important in phosphorus solubilization.

4. Other aspects of shifting cultivation

a. Purpose of burn

It is often claimed that shifting cultivators burn the slash after cutting in order to fertilize the soil by ash input (Nye and Greenland, 1960). The San Carlos experiment showed this probably is true, but it is the decomposing unburned slash on top of the soil that appears to sustain the fertility of the soil over the three-year cultivation cycle, after the ash has dissolved. Intense burns result in a higher amount of initial nutrient input into the soil, but they also result in an earlier depletion of the slash remaining after the burn.

There are several other reasons for the burning of slash in the shifting cultivation cycle. One is to kill tree stumps so that sprouting is less. Stump sprouts, if they are not killed, rapidly overtop and shade crop plants. Even after burning at San Carlos, not all stumps were killed and it was neessary to periodically cut the regrowth.

Another reason for burning is simply to permit easier access to the plot for planting. Branches and twigs in an unburned plot present formidable obstacles to passage.

Sterilization of the soil also has been mentioned as a reason for burning (Nye and Greenland 1960). Experimental burning at San Carlos did reduce the pool of viable seed in the soil from 752 to 157 per square meter (Uhl, 1980). However, the elimination of microbes, if it does occur, is not long lasting. Potential denitrification in the cultivated plot was 50 kg/ha/yr compared to 4 kg/ha/yr in the control site (Jordan *et al.*, 1983).

b. Energetic efficiency of shifting cultivation

Uhl and Murphy (1981) estimated the energy output to energy input ratio of shifting cultivation at San Carlos. Energy output was the caloric content of processed roots, and input was the caloric value of all field and processing labor, including the removal of cyanogenic compounds. They calculated a value of 13.9:1, a value much higher than for mechanized agriculture (2.82:1, Pimentel *et al.*, 1973), but lower than that found in other studies of shifting cultivation (Uhl and Murphy, 1981).

The 13.9:1 ratio was based on a total energy output of 8.4×10^6 kcal/ha and a total input of 6.06×10^5 kcal/ha. However, if we take the viewpoint that agriculture was possible only because nutrients were supplied by burning the forest, a very different picture emerges. The caloric value of the leaves in the forest was about 5.3 kcal/gram (Table 3.3). Wood values may be about 5 per cent higher (Jordan, 1971), but for this illustration we will assume the 5.3 value for the entire standing crop, which on the experimental site was 324.6 tons per hectare (Table C.2.1). The total caloric value of the forest, then, was 1720×10^6 kcal/ha, and the total energy inputs to the cycle of slash, burn and cultivate were 1720.6×10^6 kcal/ha. The output–input ratio then was only .005:1. Thus when the services of nature are included, shifting cultivation becomes very energy inefficient. Those services are gathering nutrients from a dispersed and unaccessable state in soil, and concentrating the nutrients in a form which can be readily utilized by crops.

D. CONCLUSION

This chapter began with the question: 'How do nutrient cycles and net primary productivity change during slash-and-burn agriculture at San

Carlos?' The experiment showed that while crop productivity declined sharply over the course of cultivation, total ecosystem productivity declined less and the productivity of weeds actually increased. The experiment also showed that while there was some loss of total nutrient stocks, especially potassium, from the ecosystem, the totals in the soil remained relatively high. Available phosphorus, however, declined, apparently as a result of declining soil pH, but also possibly as a result of the decreasing input of organic compounds into the soil.

Net primary productivity and nutrient dynamics in the experimental plot continued to change following abandonment of cultivation and during the ensuing secondary succession. How they changed is the subject of the next chapter.

CHAPTER 6

PRODUCTIVITY AND NUTRIENT DYNAMICS DURING SECONDARY SUCCESSION

Observations of rapid decline in crop productivity in the studies cited by Richards (1952) and quoted in the Introduction to this book led to the hypothesis that slash-and-burn agriculture resulted in nutrient-depletion of the soil. The slash-and-burn experiment at San Carlos showed that crop yields did indeed decline rapidly, but total stocks of nutrients in the soil did not appear to be seriously depleted. It was the availability of the nutrients, especially phosphorus, which seemed to be the critical factor.

What does this mean in terms of the productive potential of the slash-and-burn site after abandonment? Will primary productivity continue to decline? It is difficult to predict from the productivity data of the site during cultivation. Crop productivity is only part of the total ecosystem productivity. At the same time that crop productivity was declining, weed (secondary successional vegetation) productivity was increasing, but not very much. Weed productivity the third year was only 24 per cent of the total net primary productivity that year (Table 5.1). The increase in weed productivity could have resulted simply from less thorough weeding during the last year of cultivation, at the same time as the total net primary productivity was declining.

The question addressed in this chapter is: does the total net primary productivity of the slash-and-burn plot continue to decline following abandonment, as might be expected due to declining availability of nutrients? Or, is the successional vegetation better adapted to declining soil nutrients than the crop vegetation, and does that increase in successional productivity observed during cultivation continue following abandonment? The question was answered by a continuation of the study of nutrient dynamics and primary productivity in the experimental plot following abandonment of cultivation.

93

A. METHODS

The same experimental plot used for the slash-and-burn experiment was studied to determine primary productivity and nutrient balance following the abandonment of cultivation in January, 1980. The same plot used as a control for the period of cultivation continued to serve this purpose for post-abandonment studies.

When cultivation ended, native successional species rapidly overtopped most of the crop plants except the cashew trees. The methods used to determine the productivity of the successional vegetation were similar to those used for the crops (Appendix C.4). Productivity in the experimental site was measured for three years following abandonment, but in the control plot only above-ground biomass increment was measured after 1980. Nutrient input and output measurements for both the control and experimental plots were continued until 1983, using the same methods as during the experimental period.

B. RESULTS

Productivity during the first three years of succession in the slash-and-burn site is compared with productivity in the control site in Table 6.1. Nutrient leaching following abandonment of cultivation is included in Figs 5.6 and 5.7, and total ecosystem stocks and balance in Figs 5.8 and 5.9.

C. DISCUSSION

1. Productivity

Only wood productivity data are available for the control plot after 1979. Wood production increased in the control plot between 1980 and 1983 by about 2.6 tons per hectare annually. The factor responsible for this increase is not known.

The productivity of successional vegetation in the experimental plot also increased between 1980 and 1983, and part of that increase may have been caused by the same factors which affected the control plot, However, the increase in experimental site productivity from about 4.1 tons per hectare in the last year of cultivation (Table 5.1) to 14.1 tons per hectare in the last year of data in Table 6.1 was a much larger increase than that shown for the wood production in the control plot. This larger increase suggests that development of the structure of the successional forest was more important that any extrinsic variable such as climate in causing the increase in productivity. The above-ground standing stock of biomass increased from 7.08 tons per hectare in 1981 to 12.82 in 1982 and 19.95 in 1983, and the corresponding stocks of root biomass were 0.65, 2.11, and 2.84 tons per hectare (Uhl, 1987).

94

Table 6.1 Net primary productivity of experimental slash-and-burn plot following abandonment of cultivation (Uhl, 1987), and of wood in control plot during the dame period.

Year	kg/ha/yr, dry wt		
	1	2	3
Site			
Experimental			
Aboveground increment	5030	5740	7130
Litter fall	2240	6010	7000
Total production	7270	11750	14120
Control			
Wood*	3273	5842	5842

*Wood production data for years 2 and 3 is the average over the period

These data clearly answer the question as to whether net primary productivity declines after abandonment of slash-and-burn agriculture and confirm that it does not decline, but rather increases dramatically. It is important to point out, though, that the length of disturbance during the experiment was only three years, and that slash-and-burn agriculture does not degrade the soil as severely as other treatments such as pasture establishment using bulldozers (Jordan, 1985). Following disturbances that arc longer and more severe than slash-and-burn agriculture, net primary productivity could very well be severely depressed.

Cut, burned, and abandoned plot

In the cut, burned and abandoned plot within the experimental site, immediately adjacent to the slash-and-burn plot, above-ground productivity also returned to the levels of the control forest within two years (Table 6.2). Productivity during the first year, in contrast to what might be expected, was lower than in the conuco during the first year after abandonment (compare Tables 6.1 and 6.2).

There are at least two possible reasons for this. First, at the time the conuco was abandoned, secondary vegatation that had become established after the last weeding was already growing vigorously. In contrast, the standing stock of living biomass was close to zero in the cut, burned and abandoned plot. Secondly, since an unusually dry period immediately

Table 6.2 Above ground plant production (kg/ha/yr) in the cut, burn, and abandon site immediately adjacent to the cultivated site (from Uhl and Jordan 1984).

	Year				
	1	*2*	*3*	*4*	*5*
Above ground increment	550	9,090	7,510	11,730	11,150
Litter fall	390	3,170	4,110	6,090	8,250
Total above ground production	940	12,260	11,620	17,820	19,400

followed the burn, seed germination in the cut, burned and abandoned plot could have been inhibited (Uhl *et al.,* 1981).

Although net primary productivity in the cut, burned and abandoned plot was lower the first year than in the abandoned conuco, by the second year the above-ground wood increment in the former was considerably higher. The difference may be related to the structure of the vegetation in the two successions (Uhl *et al.,* 1982). On cultivated sites, forbs and grasses increase in abundance because their life cycles are shorter than the interval between weedings. Successional woody species are weeded out before they can produce seed locally, and after several weedings the seed bank that gives rise to this group is greatly reduced. In areas that are disturbed, but not cultivated and weeded, grasses and forbs are less important. Trees are dominant and, because of their larger structure, they have greater productive potential.

By the fourth year following abandonment, the above-ground wood productivity in the cut, burned and abandoned plot was considerably higher than that in the control forest (compare Tables 6.1 and 6.2). Relatively high net primary productivity may be characteristic of secondary successional vegetation where a high proportion of the photosynthate goes into biomass production. In mature forests, more may go into maintenance of the existing biomass (Odum 1969).

2. Nutrient stocks and dynamics

By the time the conuco was abandoned in 1980, rates of nutrient leaching had decreased to almost the levels of the control forest, although calcium and magnesium remained slightly elevated (Figs. 5.6 and 5.7). Net cumulative leaching after 1979 was very close to zero, and leaching had negligible influence on the stocks of nutrients in the ecosystem (Figs. 5.8 and 5.9). The segment of these figures showing losses due to harvesting stops at the end of

1979, because that is when harvesting was discontinued. The portion of the nitrogen figure showing net nitrogen flux also stops after 1979, because no further measurements were made.

From the time of the burn, through to the end of 1979, decreases in the total ecosystem stocks of nutrients closely paralleled cumulative leaching and harvest losses. However, in 1980, beginning at the time when cultivation was abandoned, there was a sharp, one-year decline in stocks of calcium, potassium, and magnesium. The decrease was not paralleled by an increase in measured cumulative leaching losses (Figs. 5.6 and 5.7). A possible explanation of this unaccounted-for loss is surface soil erosion, which was not systematically measured. Through 1979, the soil surface was pretty much covered with the decomposing remains of the root and humus mat of the original forest. This seemed to form small impediments to water-flow throughout the conuco, and thereby prevent serious erosion. However, by the end of 1979, the mat was almost completely decomposed and evidence of soil erosion was more apparent. For example, Fig. 6.1 shows a lysimeter (soil water collector) in 1980 that was installed beneath the root mat in the pre-burn forest in 1975. Considerable erosion is evident. After 1980, the

Fig. 6.1. Photograph taken in 1980 of lysimeter that had been installed beneath the root mat of the experimental plot in 1975, before the plot was cut and burned. When installed, the top of the trough was level with the top of the mineral soil. After the root mat decomposed, rainfall eroded the soil around the lysimeter.

unaccounted for losses of nutrient cations stopped. This occurred at about the same time that the roots of the secondary successional vegetation formed a network on top of the mineral soil, and litter cover was increasing.

From early 1981 onward, ecosystem stocks of calcium, potassium, and magnesium began to gradually increase. Soil stocks changed little after 1981, but there was a tendency for a continuing decline in calcium. Increasing stocks of nutrients in the ecosystems during succession means that the ecosystem is aggrading. An increase in nutrient stocks in biomass during succession was suggested to be a general phenomena by Vitousek and Reiners (1975). However, the increase in nutrient stocks in the biomass during succession at San Carlos may come in part through a continued depletion of stocks in the soil. Total stocks of nutrients in the soil probably will continue to decline until levels are approximately the same as in the pre-cut forest. This should occur at about the same time that biomass in the recovery forest equals that in the undisturbed forest.

3. Differences between native and wild species

The data from the experimental plot at San Carlos showed that crop productivity declined during cultivation, and that the most likely cause of the decline was nutrient limitation. In contrast, data showed that productivity of successional vegetation increased during the cultivation period. During this period, some of the increase during cultivation could have resulted from a decrease in the intensity of weeding. However, the dramatic increase in net primary productivity in the experimental plot following abandonment suggests that the successional vegetation is better adapted to obtaining the scarce nutrients than the crop plants.

Adaptations of the mature forest to the nutrient-poor conditions at San Carlos have already been discussed in Chapter III. Some of the same adaptations in the mature forest also appear in the successional vegetation. In this section, characteristics of crop plants and native successional vetetation are compared to determine how the wild vegetation apparently is better able to succeed under lowered conditions of nutrient availability.

a. Root characterisics

Perhaps the most conspicuous difference between crop and native vegetation is a larger root biomass in the wild species. During the first crop of yuca, only 145 kg per hectare of non-tuberous roots developed, and stocks were less in subsequent years (Uhl and Murphy, 1981). The root biomass of weeds during the cultivation period was not determined. However, during the first three years of successsion following the abandonment of cultivation, root biomass stocks went from 650 to 2840 kg per hectare (Uhl, 1987). Eventually, if succession proceeds without disturbance, root biomass should reach the value of the mature forest on Oxisol, where the value was close to 56,000 kg per hectare (Table C.2.1).

The root/shoot ratio indicates the amount of root available to support a unit of shoot, and thus is a good index of adaptation to nutrient-poor conditions. The ratio of fine roots (not tubers) of yuca to above ground biomass was about 0.06. The first three years of succession following abandonment, the root/shoot ratios of the wild vegetation were 0.09, 0.14, and 0.16 respectively (Uhl, 1987). In the mature forest on Oxisol, the ratio was 0.21 (Table C.2.1).

Larger roots, and more roots in relation to shoots, seems to be one reason for the relative success of successional species over at least some crop species such as yuca.

b. Life-span

Later successional species are longer-lived than crop species, and are able to take up nutrients during periods of abundance and store them for use in times of deficiency (Chapin, 1980). This might have been a factor at San Carlos after two or three years following abandonment, but many early successional species have life-spans shorter than those of the principal crops at San Carlos.

c. Root uptake kinetics

Under conditions of high soil fertility, plants that are able to take up nutrients rapidly often have a competitive advantage. Crop plants are often bred to produce high yields under conditions of high soil fertility. However, when soil fertility is low, plants that take up nutrients slowly often have an advantage, because species or varieties that require large supplies of nutrients function poorly or die (Chapin, 1980).

Two measures of the nutrient absorptive capacity of roots are V_{max}, the maximum rate of element uptake, and K_m, the element concentration at which half the maximum uptake rate occurs. Relatively high values indicate high rates of nutrient uptake. Haines *et al.*, (unpub.) measured these parameters for phosphorus and found that K_m was higher in yuca, the principal crop, and V_{max} was higher in yuca and in two early successional species than in later successional species (Table 6.3). This suggests that low root uptake kinetics are a factor influencing succession at San Carlos.

d. Nutrient concentrations

Low concentration of nutrients in tissue of mature forest trees may be an indication of efficient use of nutrients, and an adaptation to nutrient stress (Chapter 3). Nutrient concentrations in late successional species in the

99

Table 6.3 Kinetic parameters for roots in a field to forest succession at San Carlos de Rio Negro. K_m is in mg P/liter and V_{max} is in mg P/g dw root/30 min. (Haines *et al.*, unpub.).

	K_m (S.E.)	V_{max} (S.E.)
Manihot esculenta – crop species	20.7 (22.9)	1.3 (0.9)
Solanum stramonifolium – early successional	22.2 (15.5)	2.5 (0.9)
Cecropia ficifolia – early successional	28.5 (12.1)	3.5 (1.1)
Vismia japurensis – mid successional	4.9 (3.6)	0.9 (0.3)
Micrandra sp. – mature forest	9.5 (6.7)	0.2 (0.1)

abandoned conuco at San Carlos were lower than in early successional and crop species (Fig. 6.2). This suggests that a higher efficiency of nutrient use may be another factor in the replacement of short-lived species by longer-lived late successional species.

e. Mycorrhizae

Mycorrhizae were also discussed in chapter 3 as an adaptation to low nutrient conditions. Since many crop plants including yuca are mycorrhizal, it is not possible to cite mycorrhizae as a factor which gives wild species an advantage at San Carlos. However, mycorrhizal activity can differ greatly, and the presence of mycorrhizae with yuca does not mean that the mycorrhizae are highly effective in obtaining nutrients. Possibly a change from vesicular-arbuscular mycorrhizae which may be prevalent in the crop species to ectomycorrhizae in the forest species may be important at San Carlos, but little data are available.

f. Tolerance of acid soils

The low pH of many tropical Oxisols results in greater solubility of iron, aluminium and manganese. High concentrations of these elements are toxic for most crop plants (Sanchez, 1976). In contrast, many native rain forest species are able to tolerate much higher concentrations. Most crop species have between 0.1 and 200 ppm aluminium in their tissues (Bowen, 1979). In the Oxisol control forest, Sprick (1979) found five species with greater than

PERCENT DEVIATION FROM THE MEAN

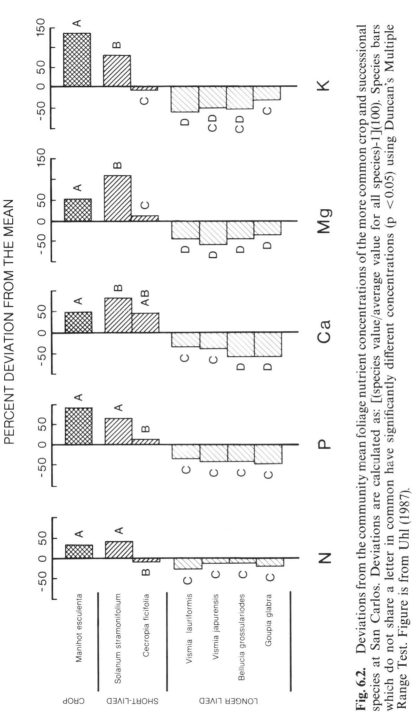

Fig. 6.2. Deviations from the community mean foliage nutrient concentrations of the more common crop and successional species at San Carlos. Deviations are calculated as: [(species value/average value for all species)-1](100). Species bars which do not share a letter in common have significantly different concentrations (p <0.05) using Duncan's Multiple Range Test. Figure is from Uhl (1987).

2000 ppm aluminum in their leaves, and in the Spodosol site she found thirteen species with more than 5000 ppm.

The mechanism by which native plants are able to avoid the aluminum toxicity which affects some crop plants is not known. Grill *et al.,* (1985) have found that some plants contain phytochelatins which bind heavy metals such as cadmium, and thereby participate in metal detoxification. They did not study binding of aluminum, so this mechanism remains to be explored.

D. CONCLUSION

Although the lack of available nutrients probably played an important role in the decline of crop production during cultivation, there was little evidence that a lack of nutrients inhibited production of successional vegetation, or will prevent recovery of the plot to mature forest. Successional vegetation appears to be better adapted than crop plants to the diminishing nutrient availability that occurs after several years of cultivation. These adaptations are the reason that rates of successional production quickly increase to levels close to, or even higher than, those of the undisturbed forest.

This result does not mean, however, that recovery to mature forest will occur following all types of disturbances. In this experiment, soil nutrient levels in the conuco at the time of abandonment were higher than those in the pre-burn forest. In contrast, if nutrients had been depleted below the level in the undisturbed forest, a rain forest resembling that before burning might not be able to regenerate. Such depletion may be occurring in some regions of the Amazon Basin. Pastures established after forest clearing in the eastern Amazon region of Brazil sometimes are grazed for ten years or more (Denevan, 1981). After the first few years, grass production decreases. Sometimes the pastures are burned to supply another pulse of readily-available nutrients. However, each burn results in a smaller pulse, and eventually the pasture is no longer feasible and the area is abandoned (Buschbacher *et al.,* 1984). Following abandonment, fire may spread into the area from adjacent areas (personal observation, Paragominas, Brazil), and the nutrient level may decrease even further. Whether such areas can ever recover is a question that has not yet been answered.

CHAPTER 7

IMPLICATIONS FOR MANAGEMENT

Alexander von Humboldt, after travels through the upper Rio Negro (Fig. 7.1) during his exploration of South America between 1799-1804 prescribed large-scale deforestation as a solution to the agricultural problems of the region. His words are (von Humboldt, p. 391):

> The banks of the Upper Guainia will be more productive when, by the destruction of the forests, the excessive humidity of the air and the soil shall be diminished. In their present state of culture, maize scarcely grows, and the tobacco which is of the finest quality, and much celebrated on the coast of Caracas, is well cultivated only on spots amid old ruins, remains of the huts of the pueblo viejo. Under a different system from that which we found existing in these countries, the Rio Negro will produce indigo, coffee, cacao, maize, and rice in abundance.

Von Humboldt was one of the first to recommend deforestation of the Amazon as a prerequisite for development. Following him, a succession of scientists, developers and politicians recommend the same thing (Sioli, 1973). Despite the repeated efforts to clear the Amazon and turn it into agricultural land, pastures or plantations, large-scale development efforts have been almost always unsuccessful. In the words of Meggars : 'The persistence of the myth of boundless productivity in spite of the ignominious failure of every large-scale effort to develop the region constitutes one of the most remarkable paradoxes of our time'.

That statement was published in 1971. Since then, there has occurred one of the largest development failures ever in the tropical regions, both in terms of economic loss and ecological destruction. That failure is of the so-called Jari project, in which approximately 12,000 sq. km of forest in the Brazilian state of Para were cut and burned, and pulp species planted for processing in

103

Fig. 7.1. Brazil-nut tree *Bertholletia excelsa* (Humboldt and Bonpland) behind the church in San Carlos in 1981. This tree was the only survivor of the group noted by Humboldt (1821).

a mill built in Japan and floated across two oceans and up the Amazon river. After 15 years in which almost a billion dollars were spent, the project was sold for a small fraction of the amount invested (Time, 1976, 1979, 1982; Kinkead 1981; Fearnside and Rankin, 1982).

The idea that deforestation is the answer to the agricultural problems of the area is perhaps one of the greatest disservices ever imposed upon the Amazon region. The results of the San Carlos project have shown that it is the forests, and the nutrients which are taken up by the trees, which are most important in making agriculture feasible, even though that agriculture requires long periods of fallow. The native forest trees are able to extract nutrients from the soil which are unavailable to crop species. When the trees are cut and burned, the ash and organic matter residue supply nutrients in a form which can be then utilized by crop species. Unburned decomposing slash also may contribute indirectly to production of crops, through mobilization of bound phosphorus. This conversion by native vegetation of unavailable nutrients into nutrients in a form which can be readily utilized by crops is the key to agriculture on soils low in nutrient availability, such as the Oxisol near San Carlos.

A. MANAGEMENT STRATEGIES FOR THE AMAZON BASIN

The last question posed in the introduction was: 'What are the implications of the work at the San Carlos for management of rain forests, both in the Rio Negro region and the tropics in general?' The best management schemes will be those in which existing soil organic matter is preserved and decomposing soil organic matter replaced by fresh inputs. This can best be accomplished by maintaining a forest structure and one way in which this can be achieved is by harvesting only select trees for wood products. Maintenance of the basic structure thus ensures that the leaf litter and falling debris from the remaining forest will sustain the soil organic matter.

Selective harvesting and silviculture to encourage the growth of desirable species were the basis of the Malayan Uniform System and the Nigerian Shelterwood System of forestry, practised in Africa and S.E. Asia during the first part of the 20th century by colonial foresters (Baur, 1964). Unfortunately, such selective harvesting no longer appears feasible, because of the economic expense of felling just a few trees per hectare, and because of the care that must be taken not to damage the remaining forest. However, strip-harvesting, similar to the corridor system practised in Africa but then abandoned (Kellog, 1963), remains a possibility. In this scheme strips, perhaps 50meters wide, are clear-cut for timber. If the terrain is hilly or mountainous, the strip is then cut on a contour. The trees on either side of the strip remain undisturbed, and roads through the middle of the strip provide access for machinery to remove logs. The advantage of narrow strip is that they simulate natural tree fall gaps in the undisturbed forest (Tosi, 1982).

One likely economic disadvantage of strip-harvesting native forests is that not all the species present may be usable. However, during the second rotation, desirable species can be planted. The cleared strip can also be cultivated for several years, planted with crop trees, or allowed to regenerate naturally. If agriculture is carried out, trees can be interspersed with crops, so that when these are abandoned the trees are already well established. It is important to recognize when selecting tree species to be planted that native species properly adapted to the naturally low levels of soil fertility will be more successful than exotics bred for rapid growth on fertile soils. Such tree species often have limitations similar to high-yield crop varieties on infertile acid soils.

One advantage of strip-cutting over large scale clear-cutting is that organic matter, such as leaf litter and falling trees from along the strip, can replenish soil organic matter lost during harvest or during cultivation of a crop within the strip. If natural regeneration is encouraged in the strip, then the trees bordering it are close enough to serve as a seed source. These same trees also are close enough to serve as a source for mycorrhizae, already shown to be important in forest regeneration. Once a new tree crop becomes established within the abandoned strip, trees up-slope from the strip can be clear-cut in another strip. If there is any nutrient leaching or soil erosion from this new up-slope strip, the rapidly regenerating down-slope strip may be able to intercept and utilize the nutrients (Jordan, 1982). The strip also may prevent the nutrients from washing into drainage streams.

The management of forests for native species should be sustainable for the same reason that shifting cultivation is sustainable – both rely on native vegetation fallow to restore the nutrients that are lost during cultivation and harvest. The native trees are capable of taking up scarce nutrients from the soil and of resupplying the soil with available nutrients through the fall of leaves and trunks, and other organic debris.

The size of a disturbed area is also important in the regeneration of forest. In both shifting cultivation and strip-harvesting, the harvested or cultivated land is never very far away from the native forest. The problems for regeneration occur when the deforested area is so large that organic debris, seeds and mycorrhizal spores do not move easily to the center of the disturbed area, either by wind or by animal vectors.

The question of commercial fertilization of soils often arises in discussions of sustained production in the Amazon region. There appear to be no serious biological or chemical obstacles to commercial fertilization of crops on soils typical of the region (Sanchez *et al.*, 1982). The problem is economic feasibility. Hauling fertilizer to the middle of the Amazon is expensive, and if the crop is subsequently to be exported it, too, must be transported to markets and seaports at further cost.

Another desirable alternative for the management of Amazonian forests is the establishment of tree plantations for fruit, cacao, rubber or other products whose harvest does not disturb the basic structure of the forest.

The proportion of total nutrient stocks removed when such products are harvested is relatively small, and, because the structure of the plantations resembles that of a forest, the recycling of nutrients and replenishment of organic matter are not seriously disrupted. This is well known by agronomists and foresters working in the Amazon (Alvim, 1978) and such crops are an important part of the economy in some areas of the region.

Sometimes there are problems in monoculture plantations with outbreaks of disease and pests because of the ease with which these can be transmitted from host to host. Plantations with valuable species planted in strips separated by barriers of native forest appears to be an effective way of dealing with this problem, at least for some species such as mahogany (Wadsworth, 1981), as long as the strips are wide enough to let in sufficient light for the young trees.

A common technique among experienced shifting cultivators in the Amazon is to plant valuable perrennial crops such as fruit trees among the early crops, so that there is a continuity of production (Denevan *et al.*, 1984). These trees contribute to the build up of soil organic matter at the same time as providing valuable products. The annual crops growing between the tree seedlings hold nutrients and prevent erosion until the trees themselves are large enough to stabilize the soil (Bruijnzeel, 1982).

In contrast to these management practices, which conserve soil organic matter as a means of sustaining the productivity of tropical ecosystems, there are other frequent practices which destroy or eliminate soil organic matter. The use of bulldozers to clear the land for agriculture or pasture is one of the most destructive management techniques possible in the wet tropics. Bulldozing not only removes the slash, which is an important source of nutrients, but also removes much of the soil organic matter, either through direct physical transport or through disturbance which hastens decomposition. The elimination of soil organic matter is undesirable not only because of its nutrient content, but also because it helps maintain good soil drainage. Soil organic matter increases soil particle aggregation, which in turn increases porosity (Swift, 1984). Bulldozing also has a direct physical impact on the soil – soil compaction which results from bulldozing destroys the well developed permeable structure of most undisturbed tropical Oxisols and Ultisols, resulting in surficial water flow and consequent erosion (Lal, 1981).

B. APPLICABILITY TO OTHER TROPICAL REGIONS

To what extent are these management recommendations valid throughout the humid tropics? The comparisons of native tropic forests (Chapter 4) showed that the adaptation of the native vegetation to nutrient deficiency is not so highly developed in regions of recent geological uplift or volcanic activity, such as parts of Central America. This suggests that nutrient deficiency should not be such a problem in these regions, and agricultural productivity should be sustainable without heavy and frequent inputs of fertilizer or organic matter. Nevertheless, studies of shifting cultivation from montane regions in Central and South America show that cropping periods seldom last for longer than three or four years before a fallow is necessary (Watters, 1971). Even in generally fertile soils, phosphorus in the soil of cleared fields may quickly become unavailable due to adsorption by amorphous iron and aluminum in volcanically derived soils, or by calcium in more basic soils (Brady, 1974; Sanchez, 1976).

Although the maximum cultivation period in these more recently formed tropical soils may not be any longer than that for Amazonian soils, the length of fallow required to build a stock of nutrients sufficient for a subsequent crop may be considerably less than on nutrient-poor Oxisols and Ultisols. There have been few studies of the build up of nutrient stocks during secondary succession on nutrient-rich tropical soils. In Costa Rica, nutrients from the cutting and burning of an eight-year-old secondary forest were sufficient to supply nutrients not only for perennial crops, but for such nutrient-demanding annuals as maize (Ewel *et al.*, 1981; Brown, 1982). A study of the mechanisms by which the native vegetation on rich tropical soils quickly build up their stocks of nutrients could provide important directions for management research.

C. SUMMARY

Each of the six questions posed in the introduction to this book has been answered chapter by chapter in each of the six chapters. The questions, and summaries of the answers, were:

1. *Are the rain forests at San Carlos under nutrient stress?* Relatively low population densities and low biomass suggest that they are. Unavailability of phosphorus appears to be the stress in the Oxisol site, and nitrogen in the Spodosol site.

2. *If the forests are under nutrient stress, how have they adapted to survive and function under these conditions?* Animal populations apparently are kept at low densities by the shortage of available nutrients. Tree populations have evolved nutrient conservation mechanisms such as a thick surficial root-mat which efficiently recycles nutrients.

3. *To what extent do these adaptations exist in other tropical forests?* The adaptations which characterize the prototype 'oligotrophic' forest type are most highly developed in so called 'heath forests' growing on Spodosols. They are least developed in forests growing on relatively rich soil, recently derived from mountain building, alluvial deposits, or volcanic activity. Forests growing on Oxisol, the most common soil type in the tropics, are somewhat closer to the oligotrophic end of the fertility gradient.

4. *How do nutrient cycles and net primary productivity change during slash and burn agriculture at San Carlos?* Following cutting and burning, soil fertility increases. During the period of cultivation, soil nutrient stocks decline

only slightly, but crop yield declines sharply. Yield decline is paralleled most closely with disappearance of above-ground organic matter, lowering of soil pH, and decline in availability of soil phosphorus. The literature suggests that products of organic matter decomposition may play a role in solubilizing phosphorus, but the exact nature of the interaction is still unclear.

5. *Following abandonment of slash-and-burn agriculture, is net primary productivity of successional vegetation affected by lack of nutrients?* Successional vegetation, like primary forest vegetation, is adapted to obtaining nutrients under conditions of soil fertility too low to support crops. Nutrient levels that exist after three years or less of cultivation do not appear to be low enough to inhibit successional productivity or recovery of the forest.

6. *What do the results imply for management of tropical forests on soils of low fertility?* The two most important results from the aspect of management are the following:

(a) Litter and other organic material on the soil surface are important for a continuing supply of nutrients and sustained crop growth. Management should aim toward maximizing soil organic matter. Practices which destroy soil organic matter should be minimized.

(b) Native trees are able to grow in soils with nutrient levels too low for crops. This means that management plans that include utilization of native tree species are more likely to be successful for sustained production.

APPENDIX A

Species Lists for the study sites near San Carlos de Rio Negro, Territorio Federal Amazonas, Venezuela. Latitude 1°56'N, longitude 67°03'W. (Previously Unpublished Lists Only)

APPENDIX A.1
Howard L. Clark and Ronald L. Liesner
Angiosperms

Species collected from the study sites near San Carlos de Rio Negro.

Herbarium Specimens Deposited in:Instituto Botanico, Caracas, Venezuela; Missouri Botanical Garden, St. Louis, Missouri, U.S.A.; New York Botanical Garden, Bronx, N.Y., U.S.A.

Acanthaceae	*Amasonia arborea* H.B.K.
	Justicia pectoralis Jacq.
	Mendoncia cardonae Leonard
	Mendoncia cf. schomburgkiana Nees
Anacardiaceae	*Anacardium occidentale* L.
	Tapirira guianensis Aublet
Annonaceae	*Anaxagorea brachycarpa* R.E. Fries
	Anaxagorea rufa Timmerman [ined.]
	Annona cf. gigantophylla (R.E. Fries) R.E. Fries
	Duguetia cauliflora R.E. Fries
	Duguetia dimorphopetala R.E. Fries
	Guatteria schomburgkiana Martius
	Tetrameranthus duckei R.E. Fries
	Unonopsis stipitata Diels
Apocynaceae	*Aspidosperma verruculosum* Muell.-Arg.
	Bonafousia
	Couma catingae Ducke
	Himatanthus
	Mandevilla
	Neocouma ternstroemiacea (Muell.-Arg.) Pierre
	Odontadenia
	Tabernaemontana
Aquifoliaceae	*Ilex laureola* Tr.

111

Araceae	*Anthurium atropurpureum* Schutes & Maguire
	Anthurium bonplandii Bunting
	Anthurium gracile (Rudge) Lindl.
	Heteropsis flexuosa (H.B.K.) Bunting
	Heteropsis spruceana Schott
	Heteropsis steyermarkii Bunting [ined.]
	Monstera adamsonii Schott
	Philodendron fragrantissimum (Hook.) Mth.
	Philodendron fraternum Schott
	Philodendron insigne?
	Philodendron cf. linnaei Kunth
	Philodendron paxianum Krause
	Philodendron remifolium R.E. Schultes
	Philodendron rubens Schott
	Philodendron cf. scandens?
	Stenospermation cf. steyermarkii Bunting
	Urospatha saggitifolia (Rudge) Schott
Araliaceae	*Didymopanax spruceanum* Seem.
Arecaceae	*Bactris balanophora* Spruce
	Bactris cf. integrifolia Wallace
	Euterpe
	Geonoma cf. maxima (Poit.) Kunth
	Iriartella setigera (Martius) H. Wendl.
	Jessenia bataua (Martius) Burrett
	Leopoldina piassaba Wallace
	Mauritia carana Wallace
	Oenocarpus bacaba Martius
Aristolochiaceae	*Aristolochia acutifolia* Duchartre
Asteraceae	*Erechtites hieracifolia* (L.) Raf. ex DC.
	Eupatorium cerasifolium (Sch.Bip.) Baker
	Melampodium camphorata (L.f.) Baker
	Mikania [sp. nov.] [H. Robinson, 1983]
	Mikania sprucei Baker
Bignoniaceae	*Arrabidaea egensis* Bur. & K. Schum.
	Callichlamys latifolia (L. Rich.) K. Schum.
	Distictella magnoliifolia (H.B.K.) Sandw.
	Distictis pulverulenta (Sandw.) A. Gentry
	Jacaranda copaia (Aublet) D. Don
	Martinella obovata (H.B.K.) Bur. & K. Schum.
	Pleonotoma dendrotricha Sandw.
	Pleonotoma jasminifolia (H.B.K.) Miers
	Schlegelia violacea (Aublet) Griseb.
Bombacaceae	*Catostemma*
	Rhodognaphalopsis humilis (Spruce ex Dcn.) Robyns
Boraginaceae	*Cordia naidophila* Johnston

112

Bromeliaceae	*Aechmea brevicollis* L.B. Smith
	Aechmea bromeliifolia (Rudge) Baker
	Aechmea corymbosa (Mart. ex Schult.) Mez
	Aechmea mertensii (Meyer) Schult. f
	Araeococcus flagelliformis Harms.
	Neoregelia eleutheropetala (Ule) L.B. Smith
	Pitcairnea rubiginosa Baker *var. rubiginosa*
	Vriesea socialis L.B. Smith
Burmanniaceae	*Apteria aphylla* (Nutt.) Barn. ex Small
	Campylosiphon purpurascens Bentham
	Dictyostega orobanchioides (Hook.) Miers var. *parviflora* (Bentham) Jonker
	Gymnosiphon divaricatus (Bentham) Benth. & Hooker
	Gymnosiphon longebracteolatus Snelders & Maas [ined.]
Burseraceae	*Protium*
	Tetragastris cf. panamense (Engl.) O. Ktze.
Caryocaraceae	*Caryocar glabrum* (Aubl.) Pers.
	Caryocar gracile Wittm.
Celastraceae	*Maytenis*
Chrysobalanaceae	*Acioa schultesii* Maguire
	Couepia bernardii Prance
	Couepia guianensis Aublet ssp. *guianensis*
	Hirtella schultesii Prance
	Hirtella ulei Pilg.
	Licania gracilipes Taub.
	Licania heteromorpha Bentham var. *heteromorpha*
	Licania latifolia Bentham
	Licania longistyla (Hook.f.) Fritsch
	Licania sprucei (Hook.f.) Fritsch
Combretaceae	*Buchenavia suaveolens* Eichler
	Buchenavia pallidovirens Cuatrecasas
Connaraceae	*Connarus*
Convolvulaceae	*Maripa paniculata* Barb.-Rodr.
Cucurbitaceae	*Cayaponia coriacea* Cogniaux
	Cayaponia pennigera Tul.
	Citrullus lanatus (Thunb.) Mats. & Nakai
	Gurania bignoniacea (P. & E.) C. Jeffrey
	Helmontia
Cyclanthaceae	*Asplundia venezuelensis* Harl.
	Cyclanthus bipartitus Poit.
Cyperaceae	*Calyptrocarya bicolor* (Pfeiff.) Koyama
	Cyperus diffusus Vahl
	Diplacrum capitatum (Willd.) Boeck.

	Diplasia karataefolia L.C. Rich.
	Hypolytrum strictum Poep. & Kunth ex Kunth
	Kyllinga pumila Michx.
	Rhynchospora pubera (Vahl) Boeck.
Dichapetalaceae	*Dichapetalum odoratum* Baill.
Dilleniaceae	*Doliocarpus savannarum* Sandw.
Dioscoreaceae	*Dioscorea*
Elaeocarpaceae	*Sloanea maroana* Steyermark
Eriocaulaceae	*Paepalanthus fasciculatus* (Rottb.) Kunth forma sphaerocephalus Herzog
Erythroxylaceae	*Erythroxylum hypoleucum* Plowman
Euphorbiaceae	*Apodandra loretensis* (Ule) Pax & Hoffm.
	Chaetocarpus
	Conceveiba guianensis Aublet
	Croton trinitatis Millsp.
	Gavarretia terminalis Baill.
	Hevea brasiliensis Muell.-Arg.
	Hevea guianensis Aublet
	Mabea
	Mabea cf. maynensis Muell.-Arg.
	Mabea cf. speciosa Muell.-Arg.
	Micrandra siphonoides Bentham
	Micrandra spruceana (Baill.) R. Schultes
	Micrandra sprucei (M.-Arg.) R. Schultes
	Pera schomburgkiana (Bentham) Muell.-Arg.
	Phyllanthus orbiculatus Rich.
	Phyllanthus stipulatus (Raf.) Webster
	Plukenetia brachybotria s.l. Muell.-Arg.
Flacourtiaceae	*Carpotroche grandiflora* Spruce
	Casearia arborea (L.C. Rich.) Urban
	Casearia javitensis H.B.K.
	Goupia glabra Reiss.
Gentianaceae	*Curtia quadrifolia* Maguire
	Lisianthus cf. chelonoides L.F.
	Pagaea plantaginifolia Maguire
Gesneriaceae	*Codonanthe*
	Codonanthopsis ulei Mansf.
	Nautilocalyx sp.
	Nautilocalyx sprucei Wiehler
Gnetaceae	*Gnetum schwackeanum* Taubert
Guttiferae	*Caraipa longipedicellata* Steyermark
	Clusia spathulaefolia Engler
	Clusia viscida Engler
	Havetiopsis

	Symphonia globulifera L.f.
Haemodoraceae	*Schiekia orinocensis* (H.B.K.) Meisn.
Hernandiaceae	*Sparattanthelium*
Heliconiaceae	*Heliconia acuminata* L. Rich.
	Heliconia juliana Barreiros
Hippocrateaceae	?
Humiriaceae	*Humiria balsamifera* (Aubl.) St. Hill.
	Saccoglottis mattogrossensis Malme
	Schistostemon cf. retusum (Ducke) Cuatr.
	Vantanea
Icacinaceae	*Discophora*
	Emmotum holosericeum Ducke
	Poraqueiba sericea Tul.
Lauraceae	*Mezilaurus sprucei* (Meissn.) Taubert ex Mez
	Ocotea costulata (Nees) Mez
Lecythidaceae	*Eschweilera bracteosa* (Poepp.) Miers
	Gustavia acuminata Mori
	Gustavia pulchra Miers

Leguminosae
[Caesalp.]

Aldina kunhardtiana? Cowan
Chamaecrista adiantifolia Spruce ex Bentham var. pteridophylla (Sandw.) Irwin & Barneby
Chamaecrista kunthiana (Schl. & Ch.) Irw. & Barneby
Dialium?
Dimorphandra pennigera Tul.
Eperua leucantha Bentham
Eperua purpurea Bentham
Heterostemon aff. cauliflorus Pittier
Heterostemon conjugatus Spruce ex Bentham
Macrolobium canaliculatum Spruce ex Bentham
Macrolobium aff. gracile Spruce ex Bentham
Macrolobium limbatum Spruce ex Bentham
Macrolobium venulosum Bentham
Senna bacillaris (L.f.) Irwin & Barneby
Swartzia cupavenensis Cowan

Leguminosae
[Mimos.]

Cedrelinga catenaeformis Ducke
Inga macrophylla H.B.K.
Inga myriantha Poepp. & Endl.
Pithecellobium cf. claviflorum Bentham
Pithecellobium laetum Bentham
Pithecellobium leucophyllum Spruce ex Bentham

Leguminosae
[Papil.]

Aeschynomene evenia Wright

115

	Clathrotropis cf. brachypetala (Tul.) Kleinh.
	Clitoria javitensis (H.B.K.) Bentham
	Dioclea elliptica Maxwell
	Humboldtiella
	Zornia latifolia Sm.
Linaceae	*Roucheria aff. punctata* Ducke
Loganiaceae	*Strychnos guianensis* (Aubl.) Martius
	Strychnos jobertiana Baillon
Loranthaceae	*Phthirusa aff. maculata* Rizz.
	Psittacanthus
Malpighiaceae	*Byrsonima wurdackii* Anderson
	Jubelina bracteosa (Gr.) Cuatr.
	Tetrapteris mucronata Cav.
Marantaceae	*Calathea acuminata* Steyermark
	Ischnosiphon polyphyllus (P. & E.) Koern.
	Monotagma capinense Huber [ined.]
	Monotagma exannulatum? Schumann
Marcgraviaceae	*Marcgravia*
Melastomataceae	*Aciotis viscosa* (Naud.) Triana
	Bellucia circumscissa Spruce ex Cogn.
	Bellucia grossularioides (L.) Triana
	Clidemia alternifolia Wurdack
	Clidemia capitellata (Bonpl.) D. Don
	Clidemia epibaterium DC.
	Clidemia heteroneura (DC.) Cogn.
	Clidemia hirta (L.) D. Don var. *tiliaefolia* (DC.) Macbride
	Clidemia japurensis DC. var. *heterobasis* (DC.) Wurdack
	Clidemia novemnervia (DC.) Triana
	Clidemia rubra sens. str. (Aublet) Martius
	Clidemia sericea D. Don
	Clidemia ulei Pilger
	Graffenrieda candelabrum Macbride
	Leandra aristigera (Naud.) Cogn.
	Loreya minor Cogn.
	Maieta guianensis Aublet
	Miconia ampla Triana
	Miconia dispar Bentham
	Miconia holosericea (L.) DC.
	Miconia longispicata Triana
	Miconia maroana Wurdack
	Miconia myriantha Bentham
	Miconia radulaefolia (Bentham) Naud.

	Mouriri uncitheca Morley & Wurdack
	Myrmidone lanceolata Cogniaux
	Myrmidone macrosperma (Mart.) Mart.
	Opisthocentra clidemioides Hook.f.
	Tococa guianensis Aublet
	Tococa hirta Berg ex Triana
	Tococa macrophysca Spruce ex Triana
	Tococa rotundifolia (Triana) Wurdack
Meliaceae	*Guarea cinnamomea?* Harms
	Guarea [sp. nov.?, Pennington, 1983]
	Guarea silvatica C.DC.
	Trichilia schomburgkii C.DC.
Menispermaceae	*Cissampelos andromorpha* DC.
	Sciadotenia sprucei Diels
Moraceae	*Brosimum utile* (H.B.K.) Pittier
	Cecropia sciadophylla Martius
	Ficus guianensis Desv. ex Ham.
	Helicostylis scabra (Macbr.) C.C. Berg
	Pourouma tomentosa Miq.
	Sorocea muriculata Miq.
Myristicaceae	*Compsoneura debilis* (A.DC.) Warb.
Myrsinaceae	*Cybianthus densiflorus* Miquel
	Cybianthus detergens Martius
	Cybianthus fulvo-pulverulentus (Mez) Agostini
Myrtaceae	*Myrcia barrensis* Berg.
	Myrcia bracteata (Rich.) DC.
	Myrsine
Nyctaginaceae	*Guapira cuspidata* (Heimerl) Lundell
	Neea obovata Spruce ex Hcimerl
Ochnaceae	*Ouratea*
	Sauvegesia erecta L.
Olacaceae	*Minquartia guianensis* Aublet
Orchidaceae	*Acacallis fimbriata* Rchb.f.
	Bifrenaria longicornis Lindl.
	Campylocentrum huebneri
	Epidendrum berryi
	Maxillaria discolor (Lada) Rchb.f.
	Maxillaria perparua Garay & Dunst.
	Octomeria complanata C. Schweinf.
	Palmorchis
	Pleurothallis
	Polycycnis vittata (Lindl.) Richb.f.
	Sigmatostalix amazonica Schlts.
	Spiranthes?

117

	Wullschlaegelia aphylla (Sw.) Reichb.f.
	Wullschlaegelia calcarata Bentham
Passifloraceae	*Passiflora auriculata* H.B.K.
	Passiflora variolata Poepp. & Endl.
Phytolaccaceae	*Phytolacca rivinoides* Kunth & Bouche
Piperaceae	*Peperomia macrostachya* (Vahl.) A. Dietr.
	Peperomia rotundifolia (L.) H.B.K.
	Piper holtii Trel. & Yuncker
	Piper perciliatum Trel. & Yuncker
	Piper sancarlosianum C.DC.
Poaceae	*Andropogon bicornis* L.
	Axonopus cappillaris (Lam.) Chase
	Homolepis aturensis (H.B.K.) Chase
	Panicum granuliferum H.B.K.
	Panicum laxum Swartz
	Panicum pilosum Swartz
	Pariana trichosticha Tutin
	Paspalum decumbens Swartz
Polygalaceae	*Securidaca cf. retusa* Bentham
Polygonaceae	*Coccoloba parimensis* Bentham
	Coccoloba striata Bentham
Potaliaceae	*Potalia amara* Aublet
Portulacaceae	*Portulaca rubricaulis* H.B.K.
Quiinaceae	*Quiina pteridophylla* (Radlk.) Pires
Rapateaceae	*Rapatea longipes* Spruce ex Korn.
	Rapatea paludosa Aublet var. *paludosa*
Rhizophoraceae	*Sterigmapetalum guianense* Steyermark
Rubiaceae	*Borreria latifolia* (Aubl.) Schum.
	Borreria ocimoides (Burm.) DC.
	Calycophyllum obovatum (Ducke) Ducke
	Coussarea brevicaulis Kr.
	Coussarea grandis Muell.-Arg.
	Coussarea leptoloba (Benth.& Hook.) M.-Arg.
	Faramea capillipes Muell.-Arg.
	Faramea corymbosa Aublet
	Faramea torquata Muell.-Arg.
	Ferdinandusa goudotiana Schum.
	Ferdinandusa uaupensis Spruce ex Schum.
	Geophila orbicularis (M.-Arg.) Steyerm. var. *orbicularis*
	Pagamea hirsuta Spruce ex Bentham
	Pagamea plicata Spruce ex Bentham
	Pagamea sessiliflora Spruce ex Bentham
	Palicourea corymbifera (M.-Arg.) Standley
	Palicourea foldatsii Steyermark

Palicourea grandiflora (H.B.K.) Standley
Palicourea longistipulata (M.-Arg.) Standley
Palicourea nitidella (M.-Arg.) Standley
Perama plantaginea (H.B.K.) Hook.f.
Psychotria adderleyi Steyermark
Psychotria blepharophylla (Standley) Steyermark
Psychotria casiquiaria Muell.-Arg.
Psychotria deflexa DC.
Psychotria humboldtiana (Cham.) Muell.-Arg.
Psychotria iodotricha Muell.-Arg.
Psychotria lupulina Muell.-Arg. ssp. *lupulina*
Psychotria medusula Muell.-Arg.
Psychotria podocephala (M.-Arg.) Standley
Psychotria poeppigiana Muell.-Arg. ssp. *barcellana*
Psychotria spadicea (Pitt.) Standl. & Steyerm.
Psychotria spiciflora Standley
Retiniphyllum concolor (Spruce ex Bentham) M.-Arg.
Retiniphyllum martianum Muell.-Arg.
Retiniphyllum pilosum (Spruce ex Benth.) M.-Arg.
Retiniphyllum truncatum Muell.-Arg.
Rudgea ayangannensis Steyermark
Rudgea berryi Steyermark
Rudgea duidae (Standley) Steyermark
Sabicea oblongifolia (Miq.) Steyermark
Sipanea pratensis Aubl. var. *dichotoma* (H.B.K.)
Steyermark *forma glabriloba* Steyermark

Sapindaceae   *Matayba cf. arborescens* (Aubl.) Radlk.
Matayba pungens (P. & E.) Radlk.
Paullinia capreolata (Aubl.) Radlk.

Sapotaceae   *Glycoxylon inophyllum* (Miquel) Ducke
Manilkara

Scrophulariaceae   *Lindernia diffusa* (L.) Wettst.
Simaroubaceae   *Picramnia platystachya* Killip & Cuatrecasas
Siparunaceae   *Siparuna cf. guianensis* Aublet
Siparuna micrantha A. DC.

Solanaceae   *Capsicum*
Markea
Schwenckia americana L.
Solanum altissimum Pittier
Solanum aff. arboreum H.B.K.
Solanum crinitum Lam.
Solanum sessiliflorum Dun.
Solanum stramonifolium Jacq.
Solanum subinerme Jacq.

Sterculiaceae	*Byttneria piersii* Cristob.
Ternstroemiaceae	*Ternstroemia*
Tiliaceae	*Apeiba*
Trigoniaceae	*Euphronia hirtelloides* Mart. & Zucc.
Triuridaceae	*Sciaphila albescens* Bentham
	Sciaphila purpurea Bentham
Verbenaceae	*Aegiphila integrifolia* (Jacq.) Jacq.
	Vitex klugii Mold.
Viscaceae	*Phoradendron*
Violaceae	*Amphirrox latifolia* Martius
	Rinorea macrocarpa (Martius) O. Kuntze
	Cissus sicyoides L.
Vochysiaceae	*Qualea esmeraldae* Standley ex Gleason
	Qualea pulcherrima Spruce ex Warm.
	Vochysia complicata Ducke
Xyridaceae	*Xyris esmeraldae* Steyermark
	Xyris savanensis Miq.
	Xyris spruceana Malme

APPENDIX A.2
J.D. Hall and R.F.C. Smith
Mammals

Species occurring in the vicinity of San Carlos de Rio Negro.

KEY: I = identified in field by J.D. Hall; T = noted by track or other sign by J.D. Hall; H = hunted or captured by local hunters; R = reported by local hunters to occur in vicinity.

Species	Evidence for occurrence
Marsupialia	
Didelphis azarae	I
D. marsupialis	I
Caluromys lanatus	I
C. philander	I
Marmosa cinerea	I
Monodelphis sp.	I
Philander opossum	I
Primata	
Cebus albifrons	I
C. apella	R
Alouatta seniculus	I
Callicebus torquatus	I

120

Aotus trivirgatus	H
Cacajao melanocephalus	H

Edentata
Priodontes giganteus	R
Dasypus novemcinctus	T
Cyclopes didactylus	H
Myrmecophaga tridactyla	R
Tamandua tetradactyla	R
Bradypus tridactylus	H
Choloepus didactylus	H

Rodentia
unknown sciurid	I
unknown sciurid	I
Proechimys cherriei	I
Echimys sp.	I
unknown echimyid	I
Dasyprocta sp.	I
Myoprocta sp	H
Agouti paca	T
Cavia sp.	R
Hydrochaerus hydrochaeris	R
Coendou prehensilis	H
Rhipidomys spp.	I
unknown cricetid	I
Neacomys sp.	I

Cetacea
Inia geoffrensis	I

Carnivora
Potos flavus	I
Bassaricyon alleni	H
Nasua nasua	T
Eira barbara	I
Galictis vittata	R
Lutra sp.	H
Leo (Panthera) onca	R
Felis pardalis	T
F. wiedii	H
F. tigrina	H
F. concolor	T
F. yagouaroundi	H
Atelocynus microtus	R

Perissodactyla
    Tapirus terrestris   T

Artiodactyla
    Tayassu pecari   T
    T. tajacu   H
    Mazama sp.   I
    Odocoileus virginianus   I

APPENDIX A.3
F. Vuilleumier
Birds

Species observed near San Carlos de Rio Negro, 17–22 December, 1980.

Cathartidae	*Cathartes melambrotus*
Accipitridae	(*Chondrohierax uncinatus?*)
	Ictinia plumbea
	Leucopternis albicollis
Charadriidae	*Charadrius collaris*
Scolopacidae	*Actitis macularia* (North American migrant)
Laridae	*Sterna superciliaris*
Columbidae	*Columba subvinacea*
	Columbina passerina
	Columbina minuta
Psittacidae	*Aratinga* (sp.)
	Pionites melanocephala
	Amazona (sp.)
Cuculidae	*Crotophaga ani*
Nyctibiidae	*Nyctibius* (?*grandis*)
Caprimulgidae	*Chordeiles acutipennis*
	Nyctidromus albicollis
	Caprimulgus (*cayennensis?*)
	Caprimulgus nigrescens
	Hydropsalis climacocerca
Apodidae	*Chaetura chapmani*
	Chaetura cinereiventris
	Reinarda squamata
Trochilidae	*Phaethornis squalidus*
	Phaethornis griseogularis
	Thalurania furcata
Ramphastidae	*Pteroglossus* (?*pluricinctus*)
	Ramphastos (?*culminatus*)
Picidae	*Piculus chrysochloros*
	Melanerpes cruentatus

Dendrocolaptidae	*Xiphorhynchus* (?*ocellatus*)
Furnariidae	*Xenops milleri*
Formicariidae	*Cymbilaimus lineatus*
	Myrmotherula brachyura
	Myrmotherula (?*menetriesii*)
	Cercomacra (?*cinerascens*)
	Hypocnemis cantator
	Gymnopithys rufigula
	Phlegopsis erythroptera
Cotingidae	*Cotinga cayana*
	Iodopleura isabellae
	Lipaugus vociferans
Pipridae	*Pipra erythrocephala*
Tyrannidae	*Tyrannus melancholicus*
	Ramphotrigon ruficauda
	Pipromorpha oleagina
Hirundinidae	*Tachycineta albiventer*
	Progne chalybea
	Atticora melanoleuca
	Hirundo rustica (North American migrant)
Troglodytidae	*Thryothorus coraya*
	Troglodytes aedon
Icteridae	*Icterus chrysocephalus*
	Icterus icterus
Parulidae	*Dendroica striata* (North American migrant)
	Coereba flaveola
Thraupidae	*Thraupis episcopus*
	Thraupis palmarum
	Ramphocelus carbo
	Tachyphonus cristatus
	Tachyphonus surinamus
Fringillidae	*Saltator maximus*
	Arremon taciturnus
	Sporophila castaneiventris
	Volatinia jacarina
	Ammodramus aurifrons

APPENDIX A.4
Kathleen Clark
Fish

Per cent by number and per cent by weight of fish families in the San Carlos catch by seven fishing methods, 1979–1981.

Family	Per cent of Catch by Number	Per cent of Catch by Weight
Characoidei (total)	(20.56)	(18.8)
Anostomidae	6.30	6.12
Characidae	3.38	1.64
Chilodidae	1.51	0.35
Ctenoluciidae	0.30	0.13
Curimatidae	1.00	0.17
Cynodontidae	1.14	0.93
Erythrinidae	5.15	7.96
Hemiodidae	0.09	0.05
Serrasalmidae	1.69	1.45
Siluroidei (total)	(50.74)	(46.44)
Ageniosidae	0.72	0.44
Auchenipteridae	15.61	5.85
Callichthyidae	0.02	0.01
Doradidae	13.23	4.72
Loricariidae	1.87	4.39
Pimelodidae	19.29	31.03
Gymnotoidei (total)	(4.76)	(5.36)
Apteronotidae	0.12	0.12
Electrophoridae	0.10	0.68
Gymnotidae	0.08	0.01
Rhamphichthyidae	4.46	4.55
Miscellaneous Groups (total)	(23.54)	(29.35)
Belonidae	0.01	0.001
Cichlidae	22.88	28.01
Potamotrygonidae	0.21	0.65
Sciaenidae	0.43	0.65
Symbranchidae	0.01	0.04

APPENDIX B
Methods for community studies
(Previously unpublished studies only)

APPENDIX B.1. Small Mammals
APPENDIX B.2. Fish
APPENDIX B.3. Decomposers

Appendix B.1. Small Mammals
J.D. Hall and R.F.C. Smith

Four grids, each 100 m square, were laid out at the corners of a plot 300 m on each side. The grids were separated from each other by 100 m. Each grid contained 25 trapping stations at 25 m intervals, and at every station there were four wire mesh live traps. One large (15 × 15 × 60 cm) trap and one small (7.5 × 7.5 × 30 cm) trap were set on the forest floor within a 2.5 m radius of each station. Two additional traps, one large and one small, were also firmly attached to the branches of a canopy tree within 5 m of each station. A combination of fresh pineapple, coconut meat, and canned sardines was used as bait. Traps were set for a 10-day period every three months from May, 1976 to May, 1978, inclusive. All animals captured were identified and the sex was determined. After being weighed, each animal was marked individually by toe-clipping and was then released at the site of capture.

The live-trapping plots were unenclosed. It was assumed that each trapping station captured all trappable animals for a radius of half the distance to the adjacent station. The area trapped was assumed to extend 12.5 m beyond the outermost traps and have a total of 6.25 ha.

Seventeen different species of mammals (10 rodents and 7 marsupials) were captured. Average rates of trapping success for all but the three rarest species was 289/36000 = 0.80 animals/100 trap nights. Density calculations excluded recaptures within a 10-day trapping period but included the first recapture in any period for an individual marked in some previous period.

Appendix B.2. Fish
K. Clark

Approximately 8,900 fishes caught by local fisherman, using a variety of traditional and modern fishing gear, were examined between 1979–1981. The fishing methods used vary widely. Fishing with hand lines is practiced throughout the year, although hook size, location, bait, and target species are variable. In general, rainy season methods exploit the movement of fishes to and within the flooded forest along the edge of rivers and streams, while dry season methods tend to involve visual location of fish.

The nature of dry season fishing has changed over the past 15 years with the introduction of spearguns and battery powered headlamps. While previously sight fishing was accomplished by bow and arrow, these implements are no longer used in San Carlos. Now fishes are hunted by day with mask and speargun and by night from canoes using spears and headlamps.

Although it is prohibited by law, the use of fish stupefactants derived from plants is still surreptiously practiced. Fish poisoning requires a group effort, and is practiced during the driest months of the year when fish are concentrated in isolated pools from which escape is difficult.

During the dry season, large catfishes are caught on trot lines. These lines are anchored near the bottom, and generally placed transversly across the river in deep pools. Trot lines are often fished at night and can be left unattended. Trot lines are not used during the rainy season because the current is sufficient to sweep them away. Rather, large catfishes are caught on hand lines, set lines, or drift lines.

Set lines with small hooks are an important rainy season fishing gear within the flooded forest. These lines are put out only in areas of shallow water (under 1 m) because the fisherman must be able to reach down from his canoe and disentangle the line from submerged vegetation.

Before the river begins to rise in March, large traps called 'cacures' are constructed along the river margin, generally in places where the current is strong and fish are naturally forced towards the shore. Cacures are built of a log scaffolding covered with a fencework of narrow palm slats. To remove fish from the trap, the fisherman dives inside the chamber with a small dip net.

In addition to the seven major fishing methods described, a number of lesser methods are also employed. These include the numerous variants of hand line fishing, as well as various traps, snares, small nets, and fishing by hand and with clubs.

Appendix B.3. Decomposers
R. Todd

Decomposition of the leaves of two common tree species on the Oxisol site, *Licania heteromorpha* and *Ocotea* sp., and two common species on the Spodosol site, *Micrandra sprucei* and *Eperua leucantha* were studied with the litter bag technique. Thirty-two bags were set out in each site in March, 1977, and sub-sets were collected over the next 1.2 years. The decay coefficient 'k' (Olson, 1963) for the Oxisol species was 0.37 and for the Spodosol species it was 0.60.

There is a problem with the use of individual species in litter bags to determine k for all the species, because of a large variation in decomposition rates between species at San Carlos (Cuevas, 1982). Further, the decomposition constant usually changes throughout the decay of an individual leaf

(Swift *et al.*, 1979). A value of k which better represents all the species simultaneously can be obtained from the equation:

$$k = L/X$$

if rate of litterfall equals rate of decomposition, and the size of the litter pool on the forest floor is constant (Olson, 1963). From Table C.2.1., decomposing leaf litter in the Oxisol root mat is 11.2 t/ha, and litter in the Spodosol is 6.5 t/ha (Herrera, 1979). From Table 4.5, litter fall in the Oxisol site is 5.87 t/ha/yr, and in the Spodosol site is 4.95 t/ha/yr. Therefore, k for the Oxisol site is 0.52 and for the Spodosol site is 0.76.

APPENDIX C
Methods for nutrient cycling and productivity and intermediate results

APPENDIX C.1.   SOILS
    a.   Physical Properties
    b.   Chemical Properties

APPENDIX C.2.   FOREST BIOMASS
    a.   Above Ground
    b.   Variability Among Oxisol Sites
    c.   Forest Floor and Below Ground
    d.   Variability Among All San Carlos Sites
    e.   Remains of Forest in Cultivated Plot

APPENDIX C.3.   FOREST PRODUCTIVITY
    a.   Above-ground Biomass Increment
    b.   Mortality, and Dynamics of Biomass Stocks
    c.   Root Productivity
    d.   Litter Fall

APPENDIX C.4.   POST-DISTURBANCE PRODUCTIVITY
    a.   Crop and Weed Productivity
    b.   Productivity Following Abandonment

APPENDIX C.5.   NUTRIENT CONCENTRATION IN BIOMASS
    a.   Forest
    b.   Crops and Successional Vegetation

APPENDIX C.6.   WATER BUDGET
    a.   Rainfall
    b.   Open Pan Evaporation
    c.   Throughfall
    d.   Stem Flow
    e.   Evaporation from Canopy
    f.   Evaporation from Forest Floor
    g.   Soil Moisture Tension
    h.   Forest Transpiration
    i.   Interflow
    j.   Runoff
    k.   Percolation through Root Mat
    l.   Crop Transpiration
    m.   Evaporation from Soil during Cultivation
    n.   Evapotranspiration following Abandonment

APPENDIX C.7.   NUTRIENT FLUXES
    a.   Litter and Tree-Fall in Forest

b. Import into, and Export from Cultivated Site
c. Nutrients in Precipitation
d. Nutrients in Soil Leachate
e. Nutrients in Throughfall and Stemflow
f. Gaseous Nitrogen Fluxes

APPENDIX C.1
Soils

a. Soil description

Profile descriptions of an Oxisol, Ultisol, and Spodosol near San Carlos are given in Tables C.1.1–C.1.3. The soil at the experimental slash-and-burn site and the control conformed to the description of the Petroferric Gibbsiothox.

Stark and Spratt (1977, Table C.2.3) found that 87% of the total root biomass of the undisturbed forest, and 89% of the fine root biomass occurs above 'Horizon 3' (Table C.1.1). The transition between 'Horizon 2' and 'Horizon 3' was assumed to mark the bottom of the soil volume having stocks of nutrients available for plant uptake.

By carefully excavating and weighing the soil from eighteen 0.25 m^2 soil pits, Stark and Spratt (1977) also determined that the average mass of mineral soil in 'Horizon 2' was 1347 t/ha. This was the soil mass assumed for the control forest. To determine the mass of this horizon for the experimental plot, ten cores were taken with a coring cylinder of known volume. Average bulk density in the plot before cutting was 1.17 g/cm^3 $\pm$ 0.10 (S.D.). Average depth of the horizon was 11.8 cm $\pm$ 13.9. Total soil mass was 1.17 × 11.9 = 13.81 g/cm^2 of forest floor, or 1381 t/ha. During the third year of cultivation, bulk density of the soil in the experimental plot was measured again, and found to be 1.23 $\pm$ 0.15, not significantly different from before treatment.

b. Chemical properties

Samples of mineral soil for chemical analysis were taken from the 30 × 50 m intensively studied portion of the slash-and-burn plot. In that plot, four transects eight meters apart were run across the width of the plot. On each transect, ten samples were taken equidistantly along each transect, for a total of forty samples for each sampling period. Samples were taken with an auger to the bottom of 'Horizon 2'. Samples were homogenized in the laboratory, so each sample represents the average concentration within this horizon. The first set of samples was taken four months before the experimental cut. Subsequent sampling initially occurred every six months but, following abandonment of cultivation, the period increased to once a year.

Table C.1.1 Soil profile description at pit 1/1, Petroferric Gibbsiothox, located on Oxisol hill, east of San Carlos (Dubroeucq and Sanchez, 1981).

Horizon	*Depth, cm.*	*Description*
1	000/015	Mat of leaf litter and fiberous roots, sharply delimited.
2	015/030	10YR 4.0/3 – Moist, with unidentifiable humic material, approximately 5%, bound with minerals and clay. Sandy clay, with fine sand, approximately 30% clay. Strongly developed crumb structure, medium to small aggregates, very clear. Numerous fine and medium horizontal roots. Evidence of strong faunal activity, such as galleries and cavities. Gradual transition to lower horizon.
3	030/070	10YR 5.0/6 – Moist, with unidentifiable humic material, approximately 1%. Blocks of material rich in iron oxide, and compact concretions and round stones, 10R 5.0/8, very numerous, hard, smooth. Clay, with fine sand, approximately 40% clay. Fine structure, friable aggregates slightly cohesive, with many faces, few roots. Moderate faunal activity, many termite galleries. Gradual transition to lower horizon.
4	070/140	7.5YR 6.0/8 – Moist, no organic material. Concretions rich in aluminum, compact irregular blocky, 10R 5.0/8, hard. Clay, with fine and coarse sand, approximately 50% clay. Fine aggregate structure, very friable, plastic, adhesive. Frequent smooth faces, thin clay covering aggregates and associated holes. Numerous pores, fine and tubular-gradual transition.
5	140/180	7.5YR 5.0/8 – Moist, many spots of oxide, 2.5YR 4.0/8, medium, irregular, connected. Frequent elements of aluminum in nodules, very thick, irregular, hard. Clay, with fine sand, approximately 40% clay. Little apparent structure. Friable aggregates, very plastic, cohesive, not adhesive.

Table C.1.2 Soil profile description at pit 1/6, Typic Paleaquult, Ultisol, east of San Carlos (Dubroeucq and Sanchez, 1981).

Horizon	*Depth, cm.*	*Description*
1	010/000	Mat of leaf litter and fiberous roots.
2	000/020	10YR 3.0/3 – Slightly moist, with undetermined humic material, approximately 5%, bound with the minerals. Silty sandy, with fine sand, approximately 80% sand. Crumb structure, thick. Fragile aggregates, grouped, not cohesive. Numerous fine pores, vacuoles and interstites.

		Numerous fine and medium roots, horizontal. High faunal activity. Distinct limit.
3	020/050	2.5Y 5.0/2 – Moist, with undetermined humic material, approximately 2%, clay sand, with fine sand, approximately 20% clay. Little structure, medium, many small aggregates. Friable aggregates, cohesive groups. Numerous fine tubular pores. Frequent medium roots. Moderate faunal activity, many termite galleries. Gradual limit.
4	050/090	2.5Y 6.0/2 – Slightly moist. Many spots of organic material, 10YR 4.0/3, large irregular, vertical, with clear limits, connected to roots. Clay, with fine sand, approximately 40% clay. Thick apparent structure. Aggregates slightly fragile, plastic, cohesive. Frequent clay skins, thin, over aggregates. Slightly porous, tubular. Frequent fine and medium roots. Moderate faunal activity. Gradual limit.
5	090/130	5Y 7.0/2 – Wet, no apparent organic material. Clay sand, with fine and coarse sand, approximately 30% clay. Structure massive, with substructure not clearly aggregated, thick. Plastic, not adhesive. Few thick tubular pores. Few fine roots.
6	130/135	Water table.

Table C.1.3 Soil profile description at pit 1/5, Ultic Placaquod, Spodosol, east of San Carlos (Dubroeucq and Sanchez, 1981).

Horizon Depth, cm. Description

1	010/000	Mat of leaf litter and fiberous roots.
2	000/020	7.5YR 3.0/2 – Slightly humid, with coarse organic matter and distinguishable humified material, approximately 5%, juxtaposed on the minerals. Sandy, with coarse and fine sand, fine crumb structure. Fragile aggregates, grouped, not cohesive. Numerous vacuoles and interstitial pores. Numerous thick and medium horizontal roots. Clear transition to lower horizon.
3	020/090	10YR 7.0/2 – Moist, many spots of organic matter, 10YR 6.0/3, medium, elongated connected to roots. Sandy, with fine and coarse sand. Structure of simple grains. Smooth. Many fine interstitial pores. Few vertical roots. Abrupt limit.
4	090/100	5YR 2.0/1 – Moist, with discernable humified material, approximately 10%, juxtaposed on minerals. Silty sand, with fine and coarse sand, approximately 80% sand. Massive structure, compact, not very porous. Few roots, medium, dividing, penetrating aggregates. Clear limit.

| 5 | 100/115 | 7.5YR 3.0/3 – Wet, frequent spots of organic matter, 5YR 2.0/1, large, elongated, with unclear limits, connected to roots, iron elements in crust, sandy, compact, hard. Clay sand, with thick sand, approximately 20% clay. Massive structure, hard, not porous. Few roots. Gradual limit. |
| 6 | 115/180 | 10YR 7.0/6 – Wet, many spots indicating reduction, 5Y 6.0/2, medium, elongated, connected to holes. Clay, with thick sand, approximately 40% clay. Little structure, thick. Very plastic, grouped, cohesive, not adhesive. Frequent thick clay skins, associated with pores. Numerous medium and thick pores. Moderate faunal activity, termite galleries. |

Calcium, potassium, and magnesium were initially extracted with a 1 N solution of ammonium acetate, but later the dilute sulfuric–hydrochloric acid extractant (Olson and Dean, 1965) was adopted. A test was carried out on thirty-two samples to standardize results. Each sample was divided in half, and one half was extracted with ammonium acetate and the other with double acid. The ratio obtained was used to correct the ammonoium acetate results. The double acid method also was used for soluble phosphorus.

Initially, nutrient cation concentrations in extraction solutions were determined by atomic absorption spectroscopy (Allen *et al.*, 1974). Later, solutions were analyzed by plasma emission spectroscopy (Jones, 1977). Total Kjeldahl nitrogen and total phosphorus were measured by colorimetric analyses (Technicon Industrial Systems, 1977) following standard acid digestion. Nutrient concentrations in the soil of the experimental plot before cutting are included in Table C.5.1.

Soil pH was determined in a 1:1 soil water mixture. Organic carbon was determined by the Walkley-Black method (Allison, 1965).

APPENDIX C.2.

Forest biomass

a. Above-ground

Forty-two trees distributed among twenty-eight species were selected for biomass determinations. Individuals were selected on the basis of their diameter at 1.5 m. Trees were chosen to obtain several individuals in each diameter class. In studies of temperate forests, individual regressions for each species are sometimes calculated. This was not possible at San Carlos, because species' diversity was too high. Tree biomass was regressed on the product of the diameter squared, height and wood density. Details of methods, allometric regressions and tree biomass have been given by Jordan and Uhl (1978). That report gave a species-weighted wood density of 0.96. Since then Saldariagga (1985) has presented a corrected wood density value of 0.71. Biomass values in Tables C.2.1 to C.2.3 are based on the corrected value.

Above-ground biomass on an area basis was determined by multiplying the biomass of the average tree in each diameter/height/size class by the number of trees in each size class per hectare. The total number of trees in each 5 cm diameter class for trees greater than 10 cm diameter was recorded and their height estimated in the one hectare control plot and the entire one hectare experimental plot before cutting. Trees between 5 and 10 cm dbh. were measured over transects totalling 4,000 m^2, and trees from 1 to 5 cm dbh. were measured over transects totalling 1,525 m^2 in the control and experimental plots. Jordan and Uhl (1978) have presented tables showing the proportion of biomass in each size class as well as the proportion of palms and standing dead trees. Putz (1983) determined liana biomass in the control forest. Biomass in the control and experimental sites is summarized in Table C.2.1.

Table C.2.1 Summary of mass, not including animals, in the control and experimental plots at San Carlos.

Component	Control Plot t/ha	Experimental Plot t/ha	
Stem bark	18.5	19.3	
Stem heartwood	71.9	75.4	
Stem sapwood	99.5	104.3	
Branch bark	4.8	5.0	
Branch heartwood	16.8	17.6	
Branch sapwood	23.3	24.4	
Twigs	5.0	5.3	
Total aboveground tree wood	239.8		251.3
Liana, wood	14.7	14.7	
Total aboveground wood	254.5	266.0	
Tree leaves	8.0	8.6	
Liana leaves	1.0	1.0	
Total leaves	9.0	9.6	
Total stem and leaves	263.5	275.6	
Aboveground roots	20.3	13.6	
Belowground roots	35.4	35.4	
Total roots	55.7	49.0	
Total standing	319.2	324.6	
Fallen trunks	15.7	15.7	
Humus in root mat	11.2	7.8	
Soil organic matter	46.5	47.6	
Total organic	392.6	395.7	
Mineral Soil	1347	1381	
Total ecosystem	1739.6	1776.7	

b. Variability in above-ground biomass among oxisol sites

In order to asses biomass variability between the control and experimental areas and other forests on Oxisol hills, trees (except palms) greater than 10 cm were recorded on two additional half-hectare plots on two nearby Oxisol hills. The site comparisons (Table C.2.2) show a range of values for above-ground biomass from 210 to 272 t/ha. The control and experimental sites were between the highest and lowest values. The values in this table are lower than those in the biomass summary (Table C.2.1) because the latter includes palms and trees less than 10 cm diameter.

c. Forest floor and below ground

Root biomass in undisturbed Oxisol forest was studied by Stark and Spratt (1977) in eighteen soil pits in sites '2' and '3' (Table C.2.3), so that the experimental and control sites would not be disturbed. Biomass of roots 'Horizon 2' and 'Horizon 3' averaged 35.35 t/ha (Table C.2.3). This value was used for below-ground root mass in both the control and experimental plots (Table C.2.1). Biomass of roots in the surficial root mat of the control forest was taken to be 20.29 t/ha, the value obtained for sites 2 and 3 (Table C.2.3). Root mat biomass in the experimental plot was estimated from forty

Table C.2.2 Above-ground living biomass of trees (non-palms) ≥ 10 cm dbh in Oxisol tierra firme sites near San Carlos.

Site	Biomass, dry weight t/ha
Control	211
Experimental	232
'2'	272
'3'	210

Table C.2.3 Summary of root biomass harvests in eighteen soil pits 0.25 m² in area and averaging 46 cm deep. Data are average root biomass in grams dry weight per m² of horizon ± 1 S.D. From Stark and Spratt (1977) and Stark (personal communication).

	Size class of roots, mm				Sum all sizes	g roots per g soil
	<6	6–10	10–20	>20		
Humus and root mat	1859 ± 210	78 ± 14	42 ± 10	50 ± 12	2029	.40
'Horizon 2'*	977 ± 97	373 ± 93	621 ± 85	860 ± 112	2831	.02
'Horizon 3'*	356 ± 43	81 ± 20	126 ± 20	141 ± 47	704	.002
Total	3192	532	789	1051	5564	

* Refer to Table C.1.1

root mat samples taken at the same points used for mineral soil sampling. Samples were taken with a soil corer having an area of $30.5\,cm^2$.

Fallen tree trunks greater than 20 cm diameter were determined by a complete survey of the control plot. The length, and top and bottom diameters of each log were measured and percent of decomposition was subjectively estimated. The biomass of each log was calculated to be the volume of the log times the average wood density times the percent decomposition. The total mass of fallen trunks in the experimental plot before cutting was assumed to be equal to that in the control plot (Table C.2.1).

In their study of the root mat, Stark and Spratt (1977) separated undecomposed fragments of dead organic matter from the roots. They found an average of 11.2 t/ha, and this value is used in Table C.2.1 for the control plot. In the experimental plot, humus in the mat (Table C.2.1) was separated from roots in the forty root mat samples taken for mat biomass determinations.

The amount of organic matter in the mineral soil was determined from Walkley-Black soil carbon analysis (Allison, 1965) of the forty sub-samples used for nutrient stocks of the pre-burn experimental forest. Soil organic matter values averaged 3.45%. This value is lower than that previously published by Jordan (1982). Initially, soil was sieved for particles greater than 0.5 mm, but later it was discovered that this sieving removed up to 35% of total soil weight. Correction factors were applied to the original data. Total soil organic matter in the control and experimental plots (Table C.2.1) was determined by the product of total soil mass and percent soil organic matter.

d. Variability in total biomass among all San Carlos sites

The tierra firme forests on Oxisol islands in the San Carlos region appear to be similar to each other in biomass. In contrast, other forest sites on different soil types are quite different. Table C.2.4 compares above-ground standing biomass and root biomass for the control and experimental sites with four other sites in the San Carlos region. Roots as a percentage of total biomass increase as total biomass decreases, with the exception of one Spodosol site dominated by *Eperua*.

e. Remains of forest in cultivated plot

The experimental burn consumed only part of the felled trunks of primary forest trees and very little of the mat of roots and humus on the soil surface. Mat samples were taken simultaneously with soil samples during the cultivation period. The biomass of roots and humus in the mat, plus soil organic matter, were determined with the same method used for the experimental forest before cutting. The remains of trunks of primary forest trees were measured at the time of conuco abandonment. The length, and

Table C.2.4 Comparisons of above and below-ground dry weight biomass at various sites in the San Carlos area. Above-ground values include lianas, except for the Ultisol site. Sites are listed in decreasing order of total biomass.

Site Vegetation	Soil	Above-ground biomass, d.w. t/ha	Total roots, d.w. t/ha	Total vegetation t/ha	Roots as a % of total vegetation	Author
Dominated by *Monopterix* sp.	Ultisol	423	42	465	9	Buschbacher 1984
Dominated by *Eperua leucantha*	Spodosol	377	106	483	22	Klinge and Herrera 1978
Mixed tierra firme forest (experimental)	Oxisol	276	49	325	15	This study
Mixed tierra firme forest (control)	Oxisol	264	58	322	18	This study
Dominated by *Micrandra sprucei*	Spodosol	192	95	287	33	Klinge and Herrera 1978
"Bana" or low Amazon caatinga	Spodosol	85	139	224	62	Klinge and Herrera 1978

top and bottom diameters of all logs were measured, and logs were subsampled for wood density determinations. Mass of logs was the product of log volume times wood density. To estimate nutrient standing stocks in roots of primary forest remaining at the time of conuco abandonment, six pits, 50 cm square and 40 cm deep were dug and roots were separated, dried and weighed.

APPENDIX C.3
Forest productivity

a. Above-ground biomass increment

The annual diameter increment of 120 trees on the Oxisol control plot and 120 on the Spodosol plot was measured between 1975 and 1981 with a dendrometer for trees greater than 10 cm, and vernier calipers for trees less than 10 cm. Because occasional replacement of screws was necessary in the dendrometer method, a method employing the use of a diameter tape precisely positioned on the tree (Jordan and Farnworth, 1982) was used for a single 1981–1983 growth increment interval. Diameter increment was converted to above-ground biomass increment with the use of the allometric equations used to predict biomass as a function of diameter, height and wood density (Jordan and Uhl, 1978). Height increment was determined as a fuction of diameter increment using Uhl's (unpublished) data on the height and diameter of 1,123 trees on the Oxisol site.

b. Mortality and dynamics of biomass stocks

To estimate the mortality of trees greater than 20 cm dbh., a yearly survey of newly fallen trees was carried out on three hectares of Oxisol forest between 1975 and 1980. One hectare was the control forest, and the other two hectares were plots '2' and '3' used for comparative biomass studies. The diameter of each dead tree was measured and its mass at the time of death was estimated using the equations for tree biomass (Jordan and Uhl, 1978). For trees between 10 and 20 cm diameter, Uhl (1982) mapped all trees on the one hectare control plot in 1975 and resurveyed the area again in 1980. All of the original trees that were missing or dead were noted. Similarly, trees 5 to 10 cm diameter were surveyed within two 10×100 m transects, and trees 1 to 5 cm diameter were surveyed within two 5×100 m transects. The net change in standing stock of biomass for each 5 cm diameter class due to the difference between production and mortality is given in Table C.3.1. The average net annual increment of 1.259 t/ha/yr is 0.5% of the total above-ground standing stock of wood (239.8, Table C.2.1).

Table C.3.1 Above-ground dry weight biomass increment, biomass loss through mortality, and net change between 1975 and 1980 on tierra firme forest on Oxisol at San Carlos.

Diameter Class cm	Biomass increment kg/ha/yr	Mortality kg/ha/yr	Δ Biomass kg/ha/yr
1–5	506	381	+125
5–10	388	335	+ 53
10–15	874	298	+576
15–20	558	323	+235
20–25	457	252	+205
25–30	367	161	+206
30–35	129	226	− 97
35–40	406	380	+ 26
40–50	311	460	−149
> 50	387	308	+ 79
Σ	4383	3124	+1259

c. Root productivity

Root growth in the surficial root mat was measured by determining the rate at which roots grew into fresh leaf and wood litter placed on the surface of the mat and into openings cleared in the root mat. Details of six different experiments have been given by Jordan and Escalante (1980) who estimated that root growth in the surficial mat was $117\,g/m^2/yr$, and in mineral soil to be $84\,g/m^2/yr$ for a total rate of root production of $201\,g/m^2/yr$.

In another experiment on the Oxisol site, root growth was measured in mesh cylinders filled with vermiculite and inserted into the soil (Cuevas, 1982). It was found that production averaged $159\,g/m^2/yr$ for cylinders with vermiculite alone, and was as high as $400\,g/m^2/yr$ for cylinders enriched with phosphate.

d. Litter fall

The rate of fall of leaves and of twigs less than 1 cm diameter was determined from monthly collections between 1975 and 1980 in the Oxisol and Spodosol sites. In each site there were forty-two plastic receptacles, each 32.5×37.7 cm with walls 20 cm high. One small hole was drilled in each corner of the receptacles to permit water drainage. The average monthly dry-weight leaf and twig litter fall is given in Table C.3.2, the lowest rates occurring between April and July, during the time of maximum rainfall.

Table C.3.2 Average monthly fall of leaves and of twigs less than 1 cm diameter on the undisturbed Oxisol and Spodosol sites at San Carlos, 1975–1980.

	Litter Fall			
	Oxisol		Spodosol	
Month	$g/m^2/mo$	$\pm S.D.$	$g/m^2/mo$	$\pm S.D.$
January	45.6	9.7	32.3	8.7
February	48.8	18.4	32.0	9.6
March	56.2	33.7	45.2	23.5
April	33.0	6.0	33.0	11.9
May	38.1	9.7	29.2	7.0
June	32.6	13.7	32.1	11.9
July	40.7	13.7	37.2	11.9
August	42.1	12.8	52.3	33.5
September	54.9	15.9	52.3	22.5
October	71.7	24.4	58.5	12.9
November	63.9	12.0	46.7	11.7
December	59.8	13.7	44.3	9.7
Σ	587.4		495.1	

APPENDIX C.4

Post-disturbance productivity

a. Crop and weed productivity

Yuca productivity was estimated by measurements of standing crop biomass, litter production and leaf consumption by chewing insects (Uhl and Murphy, 1981). Leaf stem and tuber biomass of all yuca crops were weighed at each harvest. In addition, twenty to thirty plants were carefully excavated at each harvest to sample fine root biomass. Subsamples of all biomass components were taken for dry-weight determination. Leaf litter production was estimated by counting leaf scars on fifty randomly located plants at each harvest period and then determining the mean weight of senesced yuca leaves. The production lost to chewing insects was estimated by removing the lowest leaf (i.e. the oldest leaf) from seventy-five randomly selected plants at the end of the first and second years. The leaf area and the area eaten were then determined using planimetry.

The numbers of pineapple plants in each of three growth stages (pre-fruiting, fruiting, post-fruiting) and the average weight of plants in each stage were combined with fruit harvest and litter collection measurements to estimate pineapple plant production for the three-year period. Litter production was estimated by bi-weekly collections of dead leaves or dead leaf parts from twenty marked plants.

Plantain standing crop biomass at two years was estimated using allometric regression determined through whole plant harvesting outside the study plot. The litter production of ten marked plantain plants was measured bi-weekly. Cashew tree above-ground production up to three years was also estimated by allometric regressions. Root weight, as a percentage of above-ground weight, was determined by destructive sampling outside the study plot. Leaf production was estimated by counting the number of leaves and leaf scars on all cashew trees on the site at three years, and then determining the mean weight of senesced leaves.

The harvest method was used to estimate weed production during the cultivation period. Twenty-seven plots, each 1×1.5 m, were randomly located in the conuco. Each time the conuco was weeded by the local farmers (at ten, sixteen, twenty-one, twenty-six and thirty-one months), all weeds in these plots were removed and weighed. Subsamples were taken, oven dried and reweighed to determine their moisture content.

b. Productivity following abandonment

The biomass of all plants less than two meters tall was estimated, one year after abandonment, by harvesting all plants (including remaining pineapple plants) present in twenty-five 1-m^2 plots. The harvested plants in each plot were grouped as forbs, grasses, successional woody plants, forest trees and vines. Plants in each group were divided into root stem and leaf fractions and were then weighed, subsampled, oven dried and reweighed. It was assumed that, after the first year, vegetation less than two meters tall made a negligible contribution to the annual biomass increment.

The biomass of all plants greater than two meters tall was estimated at one, two and three years after abandonment by a complete survey of all stems in the 1500-m^2 intensive study plot. The diameter and height of the trees were converted to stem and leaf biomass using regression equations developed from trees harvested outside the study plot. The root biomass was estimated from the root/shoot ratios of the harvested trees.

Litter production was measured by harvesting sixteen randomly located 2×2 m permanent plots at weekly intervals. The total yearly net primary productivity was the sum of litter production and net annual biomass increment of the standing crop of vegetation.

APPENDIX C.5

Nutrient concentration in biomass

a. Forest

At the time the forty-two trees were cut for biomass determination, subsamples were taken for nutrient analysis. Cross sectional samples of the trunk near the base and at the first branch were subsampled for bark, sapwood and heartwood. Two cross sections of one large branch from each tree were similarly subsampled. Approximately one kilogram of leaves and one kilogram of twigs (branches above lowermost leaf) also were collected by sampling approximately equal amounts from the top, middle and lower parts of the canopy. Initially, three subsamples for nutrient analysis were taken from each tissue type from each individual tree. After the coefficient of variation between replicate analysis was shown to be small (0.07), the number of subsamples analyzed was reduced to one per tissue per individual.

For roots, twelve subsamples were taken from the roots from each of the eighteen soil pits analyzed by Stark and Spratt (1977). For humus in the root mat, the cores taken from the root mat of the experimental plot were used. The roots were individually removed from each sample, and the remaining material, ranging from partially decomposed leaves to fine humus, was weighed and sub-sampled. For samples of decomposing logs, subsamples were taken from every fifth log encountered during the survey of fallen log biomass.

All samples were oven dried and then ground in a steel mill. Ground samples were dried again, and then 50 mg were digested at 400°C in a mixture of $HCLO_4$ and H_2SO_4 in the presence of vanadium pentoxide. Samples of standard *Brassica oleracea* (Bowen, 1969) were digested and analyzed in parallel with field samples and served as internal standards. Calcium, potassium and magnesium concentrations were determined by flame atomic absorption (Allen *et al.*, 1974). Phosphorus was determined by the molybdenum blue colorimetric procedure adapted for the Technicon AutoAnalyzer (Allen *et al.*, 1974). Total Kjeldahl nitrogen was determined by standard colorimetric procedures adapted for the Technicon AutoAnalyzer (Bremner 1965, Allen *et al.*, 1974).

Concentrations in each compartment (Table C.5.1) were multiplied by the biomass of compartments in the control and pre-cut experimental site (Table C.2.1) to give total stocks of nutrients.

b. Crops and successional vegetation

Twice during the cultivation period, ten samples each of leaf, stem, root and fruit of each crop species were taken to determine nutrient concentrations. The nutrient concentrations in conuco weeds also were measured. Weeds were separated into forb, grass, vine, successional woody and primary forest groups. Five samples each of leaf, stem and root tissue were taken from each plant group for nutrient analysis.

Table C.5.1 Concentrations of nutrients in the components of primary forest on Oxisol at San Carlos.

| Component | *mg/g dry weight $\pm$ 1 S.D.* | | | | |
	Calcium	*Potassium*	*Magnesium*	*Phosphorus*	*Nitrogen*
Leaves	1.33 ± 0.77	4.36 ± 1.14	0.71 ± 0.25	0.59 ± 0.60	18.46 ± 4.97
Stem bark	3.46 ± 2.41	1.89 ± 1.52	0.47 ± 0.20	0.29 ± 0.36	10.34 ± 5.38
Stem heartwood	0.39 ± 0.26	0.52 ± 0.45	0.16 ± 0.13	0.04 ± 0.02	1.49 ± 1.74
Stem sapwood	0.51 ± 0.28	0.59 ± 0.51	0.14 ± 0.20	0.09 ± 0.04	2.61 ± 1.41
Branch bark	3.54 ± 2.51	2.07 ± 1.39	0.36 ± 0.22	0.15 ± 0.08	10.50 ± 5.07
Branch heartwood	0.42 ± 0.26	0.90 ± 0.81	0.34 ± 0.23	0.11 ± 0.05	2.31 ± 1.21
Branch sapwood	0.44 ± 0.28	1.20 ± 2.04	0.15 ± 0.06	0.14 ± 0.06	2.70 ± 3.39
Twigs	1.52 ± 1.37	2.60 ± 1.66	0.39 ± 0.38	0.36 ± 0.39	7.33 ± 3.65
Roots	0.94†	0.89†	0.24†	0.35†	11.13†
Fallen trunks	0.31 ± 0.09	$0.16 \pm 0.0.05$	0.12 ± 0.05	0.05 ± 0.02	6.04 ± 0.81
Humus	0.45 ± 0.37	1.09 ± 1.10	0.25 ± 0.19	0.27 ± 0.13	14.20 ± 4.07
Soil – extractable ($\times 10^{-3}$)*	$5.15 \pm .83$	16.89 ± 1.96	3.84 ± 0.26	–	–
Soil – total	–	–	–	0.18 ± 0.08	1.26 ± 0.29

† Values are averages of the average concentrations in four diameter classes.
* For example, calcium is 5.15 mg/kg.

In the cut, burned and abandoned section of the experimental plot, Uhl and Jordan (1984) determined the nutrient concentrations in the five most common tree species. Approximately 500 g each of leaves, bark and wood were taken from six to eight individuals of each species. Roots were collected from throughout three pits and pooled to give three composite samples. Chemical analysis of the crop and successional vegetation was the same as for the primary forest.

APPENDIX C.6

Water budget

Table C.6.1 lists the water budget parameters determined and the periods of time during which the fluxes were measured.

MEASUREMENTS AT METEOROLOGY STATION

a. Rainfall

During the first two years of the project, rainfall was collected with a funnel mounted above the canopy of the control plot. Water was led down through a plastic tube to a standard Lambrecht rainfall recorder. Occasional

Table C.6.1 Water budget parameters determined during the San Carlos project.

Parameter	Time Period
Meteorological Station	
Rainfall	8 years
Open pan evaporation	8 years
Control Forest	
Throughfall	2 years
Stem flow	1 year
Evaporation from canopy	calculated
Evaporation from forest floor	8 intervals of 1 day
Soil moisture tension	10 months
Transpiration	13 intervals
Interflow	1 year
Runoff	calculated
Percolation through mat	calculated
Experimental Site	
Transpiration (crops)	3 intervals
Evaporation from soil (cultivated)	2 years
Transpiration (successional vegetation)	1 month

Table C.6.2 Percentage of time that soil water suction potentials at three depths in the Oxisol are within six tension ranges (from Franco and Dezzeo, 1982).

Range of Tension		Percentage of the time that soils at each depth are at tensions shown in left hand columns		
(mm Hg)	bars	20 cm	45 cm	75 cm
<0	<0	0	0	2
0–20	0–.03	0	1	3
20–50	.03–.07	26	41	58
50–100	.07–.13	47	40	33
100–330	.13–.44	20	18	4
330–1000	.44–1.33	7	0	0
		100%	100%	100%

problems were encountered with monkeys disturbing the funnels. To resolve this, a study was made to determine if preciptation records at the San Carlos weather station of the Venezuelan meteorology service could be satisfactorily used instead of records collected at the study plots. Rainfall data collected at the control plot was correlated with that collected by the Venezuelan Meteorology Service at the San Carlos station. On a monthly basis, there was no statistically significant difference between the values from the two stations when the funnels remained undisturbed (Heuveldop, 1978). Comparison of the control plot records and the weather station records for the first two years indicated that portions of some storms were missed by the control plot recorder. Therefore, precipitation data used here are taken only from the meteorology station records. Total precipitation during the experiment, from September 15, 1975 until the end of June, 1983, was 27,100 mm.

b. Open pan evaporation

Daily open pan evaporation was routinely determined at the meteorology station in San Carlos. Measurements were made with a standard 'Pan A' evaporation pan during the same interval as precipitation.

MEASUREMENTS IN CONTROL PLOT

c. Throughfall

Throughfall was collected in four, 6 m long, cut open PVC tubes which provided a total collecting area of 3.5 m^2. The water flowed down the slightly inclined tubes into a drum with 250 litres capacity, and water level was measured with an OTT water level recorder. In addition, throughfall was collected at the forest floor in twenty open-top tins, 25 cm top diameter, in order to obtain variation coefficients.

d. Stem flow

Stem flow was measured on nineteen trees in the Oxisol site and nineteen trees in the Spodosol site. Trees were fitted with moulded polyurethane collars (Likens and Eaton, 1970). Water was conducted from the collars into collection barrels through plastic tubing. Water in the barrels was measured weekly or after major storms for one year, with the use of a calibrated dip-stick. A regression of stemflow per tree per year on tree diameter was calculated for both sites. Stem flow as a function of diameter did not differ significantly between the Oxisol and Spodosol sites, so the data were pooled. Total stem flow per hectare on the Oxisol site was calculated by multiplying stem flow per size class of tree by the number of trees in each size class per hectare.

e. Evaporation from canopy

Stem flow plus throughfall accounted for 95% of the measured incoming precipitation. It was assumed that, on an annual basis, the remaining 5% of precipitation was intercepted and evaporated from the canopy.

f. Evaporation of water from the forest floor

Was determined gravimetrically using twenty open top cans, 6 cm diameter and 13 cm deep. A small hole was punched in the bottom of each can to permit the drainage of free water. Cans were partially filled with inorganic soil material from the study site, then a cylinder of organic surface material was cut and inserted on top of the soil to simulate the natural soil cover. Holes were dug in the study site with a 6 cm diameter bucket auger so that, when in place, the cans were flush with the soil surface. Early in the morning, the cans were weighed to the nearest tenth of a gram in the field laboratory, then covered and transported to the field sites where they were inserted in the holes. They were left until evening, when they were removed and brought to the laboratory for immediate reweighing. On days when there was rain, the experiment was abandoned. Evaporation from the soil during eight one-day determinations averaged 0.1 mm per day. Original data are given by Jordan and Heuveldop (1981).

g. Soil moisture tension

Estimates of monthly transpiration, described in the next section, assume that soil water does not limit transpiration in the Oxisol site. The data of Franco and Dezzeo (1982) are used here to justify that assumption. They installed three porous cup tensiometers at each of three depths (Table C.6.2) in the Oxisol site and made daily readings from November, 1981, until the end of September, 1982. Their readings were in terms of millimetres of mercury. The values in Table C.6.2 have been converted to bars (1 bar = 750.12 mm of mercury) because soil water tension frequently is expressed in this way.

The wilting coefficient for plants occurs at about fifteen bars, and the transition from rapidly available water to slowly available water occurs at about one bar under most conditions (Brady, 1974). Soil moisture tensions on the Oxisol site were always below 0.44 bars at 45 and 75 cm depth and only rarely exceeded one bar at the 20 cm depth (Table C.6.2). These data suggest that availability of water in soil to trees rarely limits transpiration on the Oxisol at San Carlos.

h. Forest transpiration

Transpiration of the forest was measured by the tritium method (Kline *et al.*, 1976). Dilution of tritium is proportional to the transpiration rate of the trees. A series of trees were injected with tritiated water, and the rate at which the tritium was diluted was determined from water samples extracted from the twigs. Transpiration in the Oxisol control forest appeared to be independent of species and directly proportional to the sapwood area of the trees (Jordan and Kline, 1977). The sapwood area of the forest was estimated by determining the depth of the sapwood-heartwood boundary as a function of diameter in the forty-two trees used for biomass determinations, and then extrapolating to the control plot based on the number of trees in each diameter class.

Transpiration for the control plot was calculated for thirteen measurement periods of two days each, following thirteen series of tritium injections over a two-year period. Second and third degree polynomial regression equations of transpiration on vapor pressure deficit during the periods of measurement were calculated (Fig. C.1). Both equations fit the observed values equally well, but at low values of vapor pressure deficit, where no transpiration measurements were available, predictions were obviously inaccurate. Therefore, the curves were extrapolated linearly to zero (dashed line in Fig. C.1). Vapor pressure deficit was then used to predict transpiration over a two-year period (Fig. C.2).

Open pan evaporation was measured daily at the San Carlos meteorological station. Open pan evaporation estimates potential evapotranspiration and can be an approximation of actual evapotranspiration as long as there is no water stress on plants. Open pan evaporation was determined for the

147

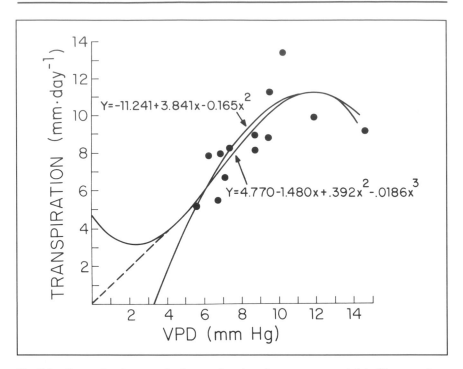

Fig. C.1. Data points for transpiration as a function of vapour pressure deficit. The regressions predict the solid curves. Curves extrapolated to zero as shown by dashed line.

same two-year period as was used for transpiration predictions by tritium. The average daily rate of evaporation of rainwater intercepted by leaves (given above) was subtracted from the open pan evaporation. The resultant rates of transpiration predicted by the open pan are compared with the results by the vapor pressure deficit in Fig. C.2. Although rates of transpiration predicted from vapor pressure deficit are always slightly higher, there is no significant difference, except during a few intervals such as January 1977. For simplicity over the eight-year period of this study, open pan evaporation was used to estimate transpiration.

i. Interflow

Interflow is the subsurface lateral drainage during and shortly after storms. To study interflow along the gradient of forest and soil types shown in Fig. 1.6, Herrera (unpublished) installed a network of forty-five steel-tube piezometers in which the depth to the water table was measured daily for one year. The network extended into the Oxisol control plot. He found that, on Oxisol hills, 'the water table is a transient phenomenon which drains

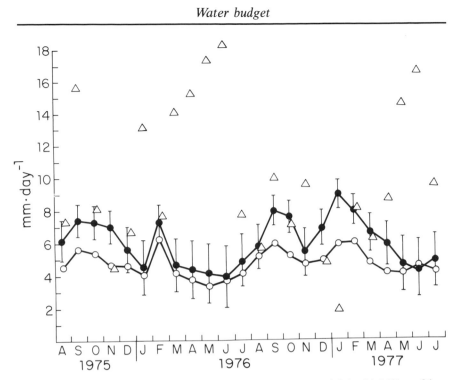

Fig. C.2. Transpiration (solid dots) predicted by vapour pressure deficit with 95% confidence interval brackets, potential transpiration (circles) estimated by evaporation pan measurements corrected for interception loss by leaves, and precipitation (triangles). Values are average mm per day for each month.

laterally in a matter of one or two hours after a heavy storm. The rooting zone is never flooded'. This suggests that on the Oxisol sites the process of interflow is important.

To estimate the proportion of water entering the soil that flows laterally as interflow in the upper soil horizon (as opposed to vertical percolation into the lower, heavier soil, before flowing laterally), tritiated water was applied to the soil surface one metre upslope from the lysimeters used for soil water collections in one pit of the experimental plot. Preparation and application procedures and analytical techniques were the same as those reported in similar studies (Kline and Jordan, 1968; Jordan *et al.*, 1970). The tritium collected in lysimeters at a 12 cm depth was four times as dilute as that collected at 40 cm. This suggests that four times as much water was flowing laterally at the 12 cm depth as at the 40 cm depth.

j. Runoff

Runoff was calculated on a monthly basis as the difference between the amount of water entering the soil and the amount lost from the soil due to evapotranspiration. Changes in the water content of the soil were not included, because over the duration of the project there was no long-term increase or decrease in soil moisture. Short-term changes were assumed to balance statistically. Water entering the soil was the sum of throughfall plus stemflow. Evapotranspiration over the eight-year study period was calculated as described above. Total runoff from the control plot between September 15, 1975 and June 30, 1983, was 13,250 mm.

k. Percolation of water through root mat

In order to quantify nutrient movement from the surficial root and humus mat into the soil, it was necessary to determine water flux through the mat as a function of throughfall. Large flat collection trays as used by Stark and Jordan (1978) were inserted under the root mat, and volumes of water moving through were measured. The attempt was not satisfactory because throughfall is so variable from point to point that no relationship could be established. Percolation of water through the root mat was calculated assuming that all the water that runs off first passes through the root mat, and the amount lost to transpirational uptake from the mat is proportional to the amount of fine roots that are in this mat.

MEASUREMENTS IN EXPERIMENTAL SITE

l. Crop transpiration

Transpiration was measured in the yuca by the tritium technique (Kline *et al.*, 1970). Transpiration rates were measured in fourteen yuca plants greater than one metre in height. Each plant was injected near its base with 50 microlitres of a 10 mCi/ml solution of tritiated water. Four plants were injected on November 25, 1977, five were injected on March 31, 1978, and five on April 3, 1978. Petiole samples for water analysis were taken hourly for two days, long enough for the tritium to move through the plants. The petioles were sealed in small plastic bags and, later, water was extracted from them by a freeze dry apparatus specially modified to capture extracted water.

To calculate transpiration of the crop, the average transpiration rate per plant was multiplied by the planting density. Yuca requires about four months to reach one metre in height. To obtain transpiration values for the first four months, we interpolated between the value for month four and zero. Transpiration was not measured in plantain, cashew or pineapple, nor in weeds, because of the small leaf area of these plants compared to yuca. This could have caused an error of about 5% in the estimate of total conuco evapotranspiration, based on the ratio of leaf area produced by yuca to that of all other plants in the conuco.

m. Evaporation from the soil during cultivation

Evaporation from the soil in the conuco was determined with the same methods used for the forest. Determinations were made on three consecutive days every two to three months during the second year of cultivation. The resulting values were correlated with evaporation rates from twelve piche tubes suspended 1 cm above the soil every three metres along a transect running diagonally across the plot. The tubes were read at 8.00 a.m. and 5.00 p.m. five days a week throughout the first two years of the experimental period. The results were extrapolated to the third year.

n. Evapotranspiration following abandonment

Evapotranspiration during secondary succession following abandonment of cultivation was estimated using the modified Penman-Monteith equation (Monteith 1965; 1973), as follows:

$$LE = \frac{s(Rn) + (dp/Ra)\ V}{s + c\ (1 + Rc/Ra)}$$

where
L = the latent heat of vaporization of water, cal/g
E = daily evapotranspiration, cm/day
s = change of saturation vapor pressure with temperature
Rn = daily net radiation flux density, cal/day
d = density of air, g/cc
p = specific heat of air
V = vapor pressure deficit
c = psychrometric constant
Ra = aerodynamic resistance of canopy
Rc = canopy resistance

The net radiation flux used in the equation was calculated from incoming radiation. Incoming direct and diffuse solar radiation was measured daily at the San Carlos weather station using a Robitsch actinograph. Net radiation was measured for a one-month period with a Fritschen shielded net radiometer. The net radiometer was placed 3 m above the vegetation and readings were recorded continuously on a strip chart for one month. A regression equation was developed to estimate net radiation at the site from incoming radiation at the weather station.

The average daytime vapour pressure deficit was calculated on a monthly basis as the difference between saturation and actual vapour pressure. These were determined from ten-year monthly averages of humidity and temperature at the meteorology station.

Canopy resistance is the quotient of stomatal resistance divided by leaf area index. Stomatal resistance of various woody species was measured by

Luvall (1981, unpublished) using a Li-Cor diffusive resistance meter. Hourly measurements were taken both adaxially and abaxially over the course of several days.

APPENDIX C.7
Nutrient fluxes

a. Litter and tree fall in undisturbed forest

Leaf and fine litter collected for productivity determinations (Appendix C.3.d) was subsampled for nutrient analysis. Samples were ground, digested and analyzed with the same methods used for nutrients in living biomass (Appendix C.5.a), except that the weight of subsamples was about 10 g.

Nutrient concentration in falling trees was not measured. It was assumed to be the same as for standing living trees (Appendix C.5.a). However, since some trees are partially decomposed when they fall, this assumption was not correct, but the error was compensated for, in as much as the nutrient loss from standing dead trees was not measured.

b. Import into and export from cultivated site

The importation of nutrients into the cultivated site was the product of the concentration of nutrients in the imported plant material (yuca stems and pineapple stocks for planting) times biomass of imported material. The exportation of nutrients was the product of the harvested material removed from the site times the biomass of removed material.

c. Nutrients in precipitation

The inputs of nutrients to the site from the atmosphere was determined using bulk precipitation collectors. The collectors were twenty narrow-mouthed, one litre polypropylene bottles. In this region of heavy rainfall, the narrow mouths of the bottles were convenient proportional samplers. Each bottle was suspended on a pole about two metres above the ground in an open area approximately one hundred metres from the forest edge and four hundred metres from the road. The mouths of the bottles were covered with a very thin layer of glass wool to prevent insects from entering. Dry fall impacted on the glass wool was assumed to be washed into the bottle by the next rain event. The bottles were emptied once a week from September 15, 1975, to June 30, 1983. Empirical correction factors were applied to concentration data to compensate for evaporation during sampling intervals. Each week, after the bottles were emptied, 50 microlitres of phenyl mercuric acetate (PMA) solution was pipetted into each collector to prevent microbial growth. Water from the field collection bottles was pooled in storage bottles at the field station. Pooled samples were taken to the laboratory in Caracas at intervals depending on the flight schedule.

Analyses of calcium, magnesium and potassium in water were done by Atomic Absorption Spectrophotometry. Nitrate plus nitrite concentrations were determined with the copper-cadmium reduction technique, ammonium with the indophenol blue method, and soluble molybdate-reactive phosphorus with the molybdenum blue technique, adapted for the Technicon AutoAnalyzer (Allen *et al.*, 1974).

A series of samples collected in 1980 were split, and half were analyzed in the Global Precipitation Network laboratory (Galloway *et al.*, 1982). Comparisons of results showed agreement for all elements except phosphate. Phosphate values from the San Carlos project laboratory were too high for aqueous samples with phosphate concentrations less than 0.1 ppm. The colorimeter used for phosphate was changed, and only post-1981 data are presented here.

d. Nutrients in soil leachate

Samples of soil water draining from the control and experimental plots were collected with 'zero-tension' type lysimeters (Jordan, 1968), so called because they collect saturated flow due to gravitational force but do not collect capillary water. Water movement due to capillary flow is in the direction from high water content to low water content. Because of the rainfall regime at San Carlos, the subsoil probably is rarely drier than the surface horizons, and nutrient loss by downward capillary flow of water probably was not important.

Lysimeters were installed in horizontal tunnels radiating from installation pits. There were six installation pits in the control plot and eight in the experimental area. Five of the pits in the experimental area were in the cultivated plot and three were in the cut, burned and abandoned area (Uhl and Jordan, 1984). From each pit, four lysimeters were installed beneath the root and humus mat, two at 12 cm and two at 40 cm. The pits were refilled after lysimeter installation.

Lysimeters were pumped dry weekly through Tygon tubing running from the buried lysimeter to the soil surface. The collection period was the same as for the precipitation samples. Samples from each lysimeter were pooled separately. Fifty microliters of PMA solution was pipetted into each pooling bottle at the start of each pooling period.

Sample analyses procedures were the same as for the precipitation samples. In addition, fifty samples were analyzed for nutrients in suspended particulate form. Concentrations were less than one per cent of dissolved concentrations, so this analysis was discontinued.

The total amount of nutrients in each soil flux was the concentration

153

determined from lysimeter samples times the volume of flux at each depth determined from the water budget (Appendix C.6). The interflow studies indicated that four times as much water flowed laterally at the 12 cm depth as at the 40 cm depth. Therefore, in the calculations of total nutrient losses due to interflow, nutrient concentrations from the 12 cm depth were weighted four times as heavily as runoff from the 40 cm depth.

To assess the influence of variations in lateral flow on estimates of leaching losses of nutrients, two sets of calculations were carried out. In one, all drainage was assumed to be lateral at 12 cm or above. In the other, all drainage was assumed to percolate vertically past 40 cm before flowing laterally. Results of the comparison between the two sets of calculations showed a difference of 9% for potassium, 5% for calcium, 4% for nitrate nitrogen, and less than 1% for magnesium, ammonium nitrogen and phosphate.

e. Nutrients in throughfall and stemflow

Throughfall and stemflow samples for nutrient analyses were collected for one year during 1975–1976. Analyses were the same as for the precipitation samples. Volumes were from the water budget (Appendix C.6). Details of collections, and results of throughfall have been presented by Jordan *et al.* (1980), and stemflow by Jordan (1978).

f. Gaseous nitrogen fluxes

Nitrogen fixation

Sample collection. In both the undisturbed Oxisol and Spodosol sites, samples were taken from leaves of eighteen species (half of the leaves with and half without epiphyllic lichens), bark of four species with and without lichens, three root mat samples and three mineral soil samples in March and July, 1977, and February, 1978. Samples were of a size to fit into 140 ml incubation jars. In the experimental conuco, leaf and stem samples of six yuca plants, six mineral soil samples, and six samples of the decomposing root mat were taken in March, and again in June. The plants were selected to give a range of sizes, and equal numbers of old, young and intermediate aged leaves were taken.

Sample analysis. Nitrogen fixation was determined by the acetylene reduction technique (Hardy *et al.*, 1968). The samples were put into 140 ml bottles whose tops were fitted with rubber serum stoppers. Exact dry weight of the samples was determined later. Ten per cent of the air in the bottles was replaced by acetylene injected into the bottles with a syringe. The samples were allowed to incubate for twenty four hours under field conditions. One millilitre of the sample atmosphere was removed and the ethylene-acetylene content was determined by gas chromotography. Blanks and controls were run in parallel with the samples. Calculations were based on a 3:1 C_2H_4:N_2 ratio. Fixation rates were multiplied by the mass of the respective compartments to give total fixation rates. Yearly fixation rates were calculated by interpolating between seasonal averages.

Denitrification

Potential denitrification rates were measured for soil and humus mat samples collected in January and July, 1979, in the undisturbed forests and the experimental conuco. The sampling procedures were the same as those for the nitrogen fixation measurements. Denitrification potentials were estimated from measurements of denitrifying activity in soil slurries during Phase I of denitrification. In the laboratory, anaerobic slurries were incubated following the procedures of Smith *et al.* (1978) and Smith and Tiedje (1979). This method generates N_2O at the maximum potential rate, but not necessarily at the rate which occurs in the field. N_2O was measured on a gas chromatograph equipped with a [63]Ni electron capture detector. The rates of N_2O generation were multiplied by the soil mass to estimate potential denitrification rates in the field plots.

APPENDIX D

Acknowledgements

The major support for the studies in this book came from the Ecosystem Studies Program of the U.S. National Science Foundation while the entire San Carlos project benefited from the support of CONICIT de Venezuela (Venezuelan Science Foundation), the UNESCO Man and the Biosphere Program, and the Organization of American States. Institutional support was provided by the Centro de Ecologia, of the Instituto Venezolano de Investigaciones Cientificas, Caracas, Venezuela, and the Institute of Ecology, University of Georgia, Athens, Georgia.

Many of the results and ideas presented in this book were a product of interactions with Drs. Ernesto Medina, Rafael Herrera, Hans Klinge, Eberhardt Brunig and Jochen Heuveldop. As both a pre-doctoral and post-doctoral student, Christopher Uhl made a large contribution to the studies in this book.

It is impossible to mention all the people who contributed to the San Carlos Project, but scientists whose work played an important role in this book are listed as authors on cited papers. Besides these scientists, special thanks are due to other colleagues whose time and effort made this book possible. Those who worked closest with me were: Gladys Escalante, Saundra Green, Pedro Maquirino, Laura Martin, Elisa Martinez, Andrea Nemeth, Margie Shedd, Joan Yantko and Steven Wooten.

The logistical support at San Carlos was organized and maintained by the resident field ecologists Jerry and Jan Hall, Howard and Kathleen Clark and Nelda Dezzeo. Student researchers who contributed to logistical support as well as to studies discussed in this book are Robert Buschbacher, Jeffrey Luvall, Juan Saldarriaga, Florencia Montagnini and Robert Sanford. The idea for the ecosystem ordinations in Chapter 4 came from Geoffrey Parker, and he also reviewed an early draft of this book.

I also give special thanks to Drs. Frank Golley and Eugene Odum for their interest in and support of the San Carlos project.

Carl F. Jordan.

157

REFERENCES

Allen, S.E., Grimshaw, H.M., Parkinson, J.A. and Quarmby C. 1974. *Chemical analysis of ecological materials.* Blackwell, England.

Allison, L.E. 1965. Organic carbon. Pages 1367–1378 in C.A. Black, ed. *Methods of soil analysis*, part 2, No. 9 in the series Agronomy. American Society of Agronomy, Madison, Wisconsin.

Alvim, P.T. 1978. Perspectivas de producao agricola na regiao Amazonica. *Interciencia* **3:** 243–249.

Anderson, A.B. 1981. White-sand vegetation of Brazilian Amazonia. *Biotropica* **13:** 199–210.

Anderson, J.P.E., and Domsch, K.H. 1975. Measurement of bacterial and fungal contributions to respiration of selected agricultural and forest soils. *Canadian Journal Microbiology* **21:** 314–322.

Anderson, J.P.E. and Domsch, K.M. 1978. A physiological method for the quantitative measurement of microbial biomass in soils. *Soil Biology and Biochemistry* **10:** 215–221.

Baur, G.N. 1964. *The ecological basis of rainforest management.* Forestry Commission, New South Wales.

Beadle, N.C.W. 1962. Soil phosphate and the delimitation of plant communities in Eastern Australia.II. *Ecology* **43:** 281–288.

Beadle, N.C.W. 1966. Soil phosphate and its role in molding segments of the Australian flora and vegetation, with special reference to xeromorphy and sclerophylly. *Ecology* **47:** 992–1007.

Beckerman, S. 1980. Fishing and hunting by the Bari' of Colombia. Pages 68–109 in R.B. Hames and K.M. Kensinger, eds. *Studies in hunting and fishing in the neotropics.* Working papers on South American Indians #2. Bennington College, Bennington, Vermont.

Beek, K.J., and Bramao, D.L. 1969. Nature and geography of South American soils. Pages 82–112 in E.J. Fittkau, J. Illies, H. Klinge, G.H. Schwabe and H. Sioli eds. *Biogeography and ecology in South America.* Junk. The Hague.

Bernhard-Reversat, F. 1975. Les cycles biogeochemiques des macroelements. Chapter 5 in G. Lemee, ed. Recherches sur l'ecosysteme de la foret subequatoriale de basse Cote-D'Ivoire. Les cycles des macroelements. *La Terre et la Vie* **29:** 229–254.

Block, W. and Banage, W.B. 1968. Population density and biomass of earthworms in some Uganda soils. *Rev. Ecol. Biol. Sol.* **5:** 515–521.

Bormann, F.H., Likens, G.E. and Mellilo, J.M. 1977. Nitrogen budget for an aggrading northern hardwood forest ecosystem. *Science* **196:** 981–983.

Bourgeois, W.W., Cole, D.W., Reikerk, H. and Gessell, S.P. 1972. *Geology and soils of comparative ecosystem study areas, Costa Rica.* Contribution No. 11. Institute of Forest Products, College of Forest Resources, University of Washington.

Bowen, H.J.M. 1969. Standard materials and intercomparisons. *Advances in Activation Analysis* **1:** 101–113.

Bowen, H.J.M. 1979. *Environmental chemistry of the elements.* Academic Press, London.

Brady, N.C. 1974. *The nature and properties of soils.* Macmillan, New York.

Bremner, J.M. 1965. Total nitrogen. Pages 1149–1178 in C.A. Black ed. *Methods of soil analysis*, Part 2, No. 9 in the series Agronomy. American Society of Agronomy, Madison, Wisconsin.

Brinkman, W.L.F., and Nascimento J.C. 1973. The effect of slash and burn agriculture on plant nutrients in the tertiary region of Central Amazonia. *Acta Amazonica* **3:** 55–61.

Brown, B. 1982. *Productivity and herbivory in high and low diversity tropical*

Brown, B. 1982. *Productivity and herbivory in high and low diversity tropical successional ecosystems in Costa Rica.* Ph.D. dissertation, University of Florida, Gainesville, Florida.

Brown, S. and Lugo A.E. 1984. Biomass of tropical forests: a new estimate based on forest volumes. *Science* 223: 1290–1293.

Bruijnzeel, L.A. 1982. *Hydrological and biogeochemical aspects of man-made forests in south-central Java,* Indonesia. Thesis. Final report, Vol. 9, Nuffic Serayu Valley Project, ITC/GUA/VU/1, the Hague, Netherlands.

Buschbacher, R., Uhl, C. and Serrao, A. 1984. Forest development following pasture use in the north of Para, Brazil. Proceedings of the first symposium on development in the humid tropics. M. Dantas ed. Belem, Brazil: *Empresa Brasileira de Pesquisa Agw Pecuria.*

Camargo, M.N., and Falesi, I.C. 1975. Soils of the central plateau and transamazonic highway of Brasil. Pages 25–45 in E. Bornemiza and A. Alvarado eds. *Soil management in tropical America.* Soil Science Dept., N.C. State University, Raleigh, N.C.

Chapin, F.S. 1980. The mineral nutrition of wild plants. *Annual Review Ecology Systematics* **11**: 233–260.

Charley, J.J., and Richards, B.N. 1983. Nutrient allocation in plant communities: mineral cycling in terrestrial ecosystems. Pages 5–45 in O.L. Lange, P.S. Nobel, C.B. Osmond and H. Ziegler eds. *Physiological plant ecology IV. Ecosystem processes: mineral cycling and man's influence.* Springer-Verlag. Berlin.

Cline, G.R., Powell, P.E., Szaniszlo, P.J. and Reid, C.P.P. 1982. Comparison of the abilities of hydroxamic, synthetic, and other natural organic acids to chelate iron and other ions in nutrient solution. *Soil Science Society of America Journal* **46**: 1158–1164.

Cole, C.V., and Heil, R.D. 1981. Phosphorus effects on terrestrial nitrogen cycling. Pages 363–374 in F.E. Clark and T. Rosswall (eds). *Ecological Bulletins* (Stockholm) **33**: 363–374.

Cole, D.W., Gessell, S.P. and Dice, S.F. 1967. Distribution and cycling of nitrogen, phosphorus, potassium, and calcium in a second growth Douglas fir ecosystem. Pages 197–232 in H.E. Young, ed. *Symposium on primary productivity and mineral cycling in natural ecosystems.* College of Life Sciences and Agriculture, University of Maine, Orono.

Cole, D.W. and Johnson, D.W. Undated. *Mineral cycling in tropical forests.* Publication number 1269. Environmental Sciences Division, Oak Ridge National Laboratory, Oak Ridge, Tennessee.

Coley, P.D. 1983. Herbivory and defensive characteristics of tree species in a lowland tropical forest. *Ecological Monographs* **53**: 209–233.

Cuevas, E. and Medina, E. 1988. Nutrient dynamics within amazonian forests II. Fine root growth, nutrient availability, and leaf litter decomposition. *Oecologia* **76**: 222–235.

Dalton, J.D., Russell, G.C. and Sieling, D.H. 1952. Effect of organic matter on phosphate availability. *Soil Science* **73**: 173–181.

Day, P.R. 1965. Particle fractionation and particle-size analysis. Pages 545–567 in C.A. Black ed. *Methods of Soil Analysis, Part 1.* No. 9 in the series Agronomy. American Society of Agronomy, Madison, Wisconsin.

Denevan, W.M. 1971. Campa subsistence in the Gran Pajonal, eastern Peru. *The geographical Review* **61**: 496–518.

Denevan, W.M. 1981. Swiddens and cattle versus forest: the imminent demise of the Amazon rain forest reexamined. Pages 25–44 in *Where have all the flowers gone? Deforestation in the third world.* Studies in third world societies No. 13. Dept. of

Anthropology, College of William and Mary, Williamsburg, Virginia.

Denevan, W.M., Treacy, J.M, Alcorn, J.B., Padoch, C., Denslow, J. and Flores Paitan, S. 1984. Indigenous agroforestry in the Peruvian Amazon: Bora Indian management of swidden fallows. *Interciencia* **9**: 346–357.

Doku, E.V. 1969. *Cassava in Ghana.* Ghana University Press. Accra.

Domalski, E.S. 1972. Selected values of heats of combustion and heats of formation of organic compounds containing the elements C, H, N, O, P, and S. *Journal Physical Chemical Reference* Data **1**: 221–237.

Dubroeveg, D. and Sanchez, V. 1981. Caracteristicas ambientales y edaficas del area muestra San Carlos de Rio Negro-Solano. Serie Informe Cientifico DGSIIA/IC/12. *Ministeria del ambiente y de los recursos naturales renovables.*

Edmisten, J. 1970. Soil studies in the El Verde rain forest. Pages H-79–H-87 in H.T. Odum ed. *A tropical rain forest.* Division of Technical Information, U.S. Atomic Energy Commission, Washington, D.C.

Ewel, J., Berish, C., Brown, B., Price, N. and Raich, J. 1981. Slash and burn impacts on a Costa Rican wet forest site. *Ecology* **62**: 816–829.

Farnworth, E.G. and Golley, F.B. 1974. *Fragile ecosystems.* Springer Verlag, N.Y.

Fearnside, P.M. and Rankin, J.M. 1982. The new Jari : risks and prospects of a major amazonian development. *Interciencia* **7**: 329–339.

Finn, J.T. 1976. Measures of ecosystem structure and function derived from analysis of flows. *Journal of Theoretical Biology* **56**: 363–380.

Finn, J.T. 1978. Cycling index: a general definition for cycling in compartment models. Pages 138–164 in D.C. Adriano and I.L. Brisbin, eds. *Environmental chemistry and cycling processes.* CONF-760429. Technical Information Center, U.S. Dept. Energy, Washington, D.C.

Fittkau, E.J., Junk, W., Klinge, H. and Sioli, H. 1975. Substrate and vegetation in the Amazon region. Berichte der Internationalen Symposien der Internationalen Vereinigung fur Vegetations Kunds Herausgegeben von Reinhold Tuxen. *Vegetation and Substrat.* Rinteln 31.3–3.4.1969.

Fleming, T.H. 1975. The role of small mammals in tropical ecosystems. Pages 269–298 in F.B. Golley, K. Petrusewicz and L. Ryszkowski, eds. *Small mammals: their productivity and population dynamics.* Cambridge University Press, Cambridge.

Forman, R.T. 1975. Canopy lichens with blue-green algae: a nitrogen source in a Colombian rain forest. *Ecology* **56**: 1176–1184.

Fox, R.L. and P.G.E. Searle. 1978. Phosphate adsorption by soils of the tropics. Pages 97–119 in M. Stelly, ed. *Diversity of soils in the tropics.* ASA Special Publication No. 34. American Society of Agronomy, Madison, Wis.

Franco, W. and Dezzeo, N. 1982. Consideraciones sobre el regimen hidrico de los principales tipos de suelos de los alrededores de San Carlos de Rio Negro. Ms presented at the Caracas symposium of the San Carlos project, *Estructura y funcion de ecosistemas forestales amazonicas del alto Rio Negro.* Instituto Venezolano de Investigaciones Cientificas, Caracas, Venezuela.

Galloway, J.N., Likens, G.E., Keene, W.C. and Miller, J.M. 1982. The composition of precipitation in remote areas of the world. *Journal of Geophysical Research* **87**: 8771–8786.

Gamble, T.N., Betlach, M.R. and Tiedje, J.M. 1977. Numerically dominant denitrifying bacteria from world soils. *Applied and Environmental Microbiology* **33**: 926–939.

Gauch, H.G. 1977. Ordiflex – a flexible computer program for four ordination techniques: weighted averages, polar ordination, principal components, and reciprocal averaging. *Ecology and Systematics*, Cornell University, Ithaca, N.Y.

Gauch, H.G., Whittaker, R.H. and Wentworth, T.R. 1977. A comparative study of reciprocal averaging and other ordination techniques. *Journal Ecology* **65**: 157–174.

Gerretsen, F.C. 1948. The influence of microorganisms on the phosphate intake by the plant. *Plant and Soil* **1:** 51–81.

Gessel, S.P., Cole, D.W., Johnson, D. and Turner J. 1977. The nutrient cycles of two Costa Rican forests. Pages 623–643 in *Actas del IV Symposium Internacional de Ecologia Tropical*, Vol. II, March 1977, University of Panama, Republic of Panama.

Gist, C.S. and Crossley, D.A., Jr. 1975. The litter arthropod community in a southern Appalachian hardwood forest: numbers, biomass and mineral element content. *American Midland Naturalist* **93:** 107–122.

Golley, F.B. 1961. Energy values of ecological materials. *Ecology* **42:** 581–584.

Golley, F.B. 1969. Caloric value of wet tropical forest vegetation. *Ecology* **50:** 517–519.

Golley, F.B. 1977. Insects as regulators of forest nutrient cycling. *Tropical Ecology* **18:** 116–123.

Golley, F.B., McGinnis, J.T., Clements, R.G., Child, G.I. and Deuver, M.J. 1975. *Mineral cycling in a tropical moist forest ecosystem*. University of Georgia Press, Athens, Georgia.

Goodland, R. and Pollard R. 1973. The Brazilian cerrado vegetation: a fertility gradient. *Journal of Ecology* **61:** 219–224.

Gosz, J.R. 1975. Nutrient budgets for undisturbed ecosystems along an elevational gradient in New Mexico. Pages 780–799 in F.G. Howell, J.B. Gentry, and M.H. Smith, eds. *Mineral cycling in southeastern ecosystems*. CONF 740513. Technical Information Center, U.S. Energy Research and Development Administration, Washington, D.C.

Goulding, M. 1981. Man and fisheries on an Amazon frontier. *Developments in hydrobiology* 4. W. Junk. The Hague.

Graustein, W.C., Cromack, K. and Sollins, P. 1977. Calcium oxalate: occurrence in soils and effect on nutrient and geochemical cycles. *Science* **198:** 1252–1254.

Grill, E., Winnacker, E.L. and Zenk, M.H. 1985. Phytochelatins: The principal heavy-metal complexing peptides of higher plants. *Science* **230:** 674–676.

Grubb, P.J. 1977. Control of forest growth and distribution on wet tropical mountains, with special reference to mineral nutrition. *Annual Review Ecology Systematics* **8:** 83–107.

Guiran, G. de 1965. Nematode associes au manioc dans le Sud du Togo. Pages 677–680 in *Compte-rendu des Travaux, Congres de la protection des cultures tropicales*, Marseilles.

Haines, B.L. 1978. Element and energy flows through colonies of the leaf-cutting ant, Atta colombica, in Panama. *Biotropica* **10:** 270–277.

Haines, B. 1983. Leaf-cutting ants bleed mineral elements out of rainforest in southern Venezuela. *Tropical Ecology* **24:** 85–93.

Haines, B., Uhl, C. and Abbot, D. Unpublished. Calcium and phosphorus uptake kinetics for selected Amazonian successional species.

Harcombe, P.A. 1977a. The influence of fertilization on some aspects of succession in a humid tropical forest. *Ecology* **58:** 1375–1383.

Harcombe, P.A. 1977b. Nutrient accumulation by vegetation during the first year of recovery of a tropical forest ecosystem. Pages 347–378 in J. Cairns, K.L. Dickson, and E.E. Herricks, eds. *Recovery and restoration of damaged ecosystems*. University of Virginia Press, Charlottesville.

Hardy, R.W., Holsten, R.D., Jackson, E.K. and Burns, R.C. 1968. The acetylene-ethylene assay for N_2 fixation: laboratory and field evaluation. *Plant Physiology* **43:** 1185–1207.

Harris, D.R. 1971. The ecology of swidden cultivation in the upper Orinoco rain forest, Venezuela. *The Geographical Review* **61:** 475–495.

Harrison, J.L. 1969. The abundance and population density of mammals in Malayan lowland forests. *Malayan Naturalist* **22**: 174–178.

Heinrichs, H., and Mayer, R. 1977. Distribution and cycling of major and trace elements in two central European forest ecosystems. *Journal Environmental Quality* **6**: 402–406.

Hermann, R.K. 1977. Growth and production of tree roots: a review. Pages 7–28 in J.K. Marshall ed. *The belowground ecosystem: a synthesis of plant-associated processes*. Range Science Dept. Science Series No. 26, Colorado State Univ., Fort Collins, Colorado.

Herrera, R. 1979. *Nutrient distribution and cycling in an Amazon caatinga forest on spodosols in southern Venezuela.* Ph.D. dissertation, University of Reading, Reading, England.

Herrera, R., Merida, T., Stark, N. and Jordan, C. 1978. Direct phosphorus transfer from leaf litter to roots. *Naturwissenschaften* **65**: 208–209.

Herrera, R. and Jordan, C.F. 1981. Nitrogen cycle in a tropical Amazonian rain forest: the caatinga of low mineral nutrient status. In F.E. Clark and T. Rosswall eds. *Terrestrial nitrogen cycles*. Ecological Bulletins (Stockholm) **33**: 493–505.

Heuveldop, J. 1978. The international Amazon MAB rainforest ecosystem pilot project at San Carlos de Rio Negro: micrometeorological studies. Pages 106–123 in S. Adisoemarto and E.F. Brunig eds. *Transaction of the Second International MAB-IUFRO Workshop on Tropical Rainforest Ecosystems Research*. Chair of World Forestry, Special Report No. 2. University of Hamburg, Hamburg-Reinbek.

Heuveldop, J. 1980. Bioklima von San Carlos de Rio Negro, Venezuela. *Amazoniana* **7**: 7–17.

Holdridge, L.R. 1967. *Life Zone Ecology*. Tropical Science Center, San Jose, Costa Rica.

Humboldt, A. von, 1821. *Personal narrative of travels to the equinoctial regions of the new continent, during the years 1799–1804*. Longman, London.

Jackson, M.L., Tyler, S.A., Willis, A.L., Bourdeau, G.A. and Pennington, R.P. 1948. Weathering sequence of clay-sized minerals in soils and sediments. *Journal of Physical and Colloidal Chemistry* **52**: 1237–1260.

Janos, D.P. 1983. Tropical mycorrhizas, nutrient cycles, and plant growth. Pages 327–345 in S.L. Sutton, T.C. Whitmore, and A.C. Chadwick eds. *Tropical rain forest: ecology and management*. Blackwell, Oxford.

Janzen, D.H. 1974. Tropical blackwater rivers, animals, and mast fruiting by the Dipterocarpaceae. *Biotropica* **6**: 69–103.

Janzen, D.H. and Schoener, T.W. 1968. Differences in insect abundance and diversity between tropical dry seasons. *Ecology* **49**: 98–110.

Jenny, H. 1980. The soil resource. *Ecological Studies* **37**. Springer Verlag, New York.

Johnson, D.W., Cole, D.W. and Gessel, S.P. 1975. Processes of nutrient transfer in a tropical rain forest. Biotropica **7**: 208–215.

Johnson, D.W., Cole, D.W., Gessel, S.P., Singer, M.J. and Minden, R.V. 1977. Carbonic acid leaching in a tropical, temperate, subalpine, and northern forest soil. *Arctic and Alpine Research* **9**: 329–343.

Johnson, P.L. and Swank, W.T. 1973. Studies of cation budgets in the Southern Appalachians on four experimental watersheds with contrasting vegetation. *Ecology* **54**: 70–80.

Jones, J.B. 1977. Elemental analysis of soil extracts and plant tissue ash by plasma emission spectroscopy. *Communications in Soil Science and Plant Analysis*: 349–365.

Jones, C.A., Cole, C.V., Sharpley, A.N. and Williams, J.R. 1984. A simplified soil and plant phosphorus model I. documentation. *Soil Science Society of America Journal* **48**: 800–805.

Jordan, C.F. 1968. A simple tension-free lysimeter. *Soil Science* **105**: 81–86.

Jordan, C.F. 1969. Derivation of leaf-area index from quality of light on the forest floor. *Ecology* **50**: 663–666.

Jordan, C.F. 1971a. Productivity of a tropical forest and its relation to a world pattern of energy storage. *Journal of Ecology* **59**: 127–142.

Jordan, C.F. 1971b. A world pattern in plant energetics. *American Scientist* **59**: 425–433.

Jordan, C.F. 1978a. *Nutrient dynamics of a tropical rain forest ecosystem and changes in the nutrient cycle due to cutting and burning.* Annual Report to U.S. National Science Foundation. Inst. Ecology, Univ. Georgia, Athens, Georgia.

Jordan, C.F. 1978b. Stemflow and nutrient transfer in a tropical rain forest. *Oikos* **31**: 257–263.

Jordan, C.F. 1982a. Amazon rain forests. *American Scientist* **70**: 394–401.

Jordan, C.F. 1982b. The nutrient balance of an Amazonian rain forest. *Ecology* **63**: 647–654.

Jordan, C.F. 1983. Productivity of tropical rain forest ecosystems and the implications for their use as future wood and energy sources. Pages 117–135 in F.B. Golley ed. *Tropical rain forest ecosystems.* Ecosystems of the World, 14A. Elsevier. Amsterdam.

Jordan, C.F. 1985. *Nutrient cycling in tropical forest ecosystems.* Wiley, Chichester.

Jordan, C.F., Koranda, J.J., Kline, J.R. and Martin, J.R. 1970. Tritium movement in a tropical ecosystem. *BioScience* **20**: 807–812.

Jordan, C.F., Kline, J.R. and Sasscer, D.S. 1972. Relative stability of mineral cycles in forest ecosystems. *American Naturalist* **106**: 237–253.

Jordan, C.F. and Kline, J.R. 1977. Transpiration of trees in a tropical rain forest. *Journal of Applied Ecology* **14**: 853–860.

Jordan, C.F. and Murphy, P.G. 1978. A latitudinal gradient of wood and litter production, and its implication regarding competition and species diversity in trees. *American Midland Naturalist* **99**: 415–434.

Jordan, C.F. and Uhl, C. 1978. Biomass of a "tierra firme" forest of the Amazon Basin. *Oecologia Plantarum* **13**: 387–400.

Jordan, C.F. and Stark, N. 1978. Retencion de nutrientes en la estera de raices de un bosque pluvial Amazonico. *Acta Científica Venezolana* **29**: 263–267.

Jordan, C.F, Todd, R.L. and Escalante, G. 1979. Nitrogen conservation in a tropical rain forest. *Oecologia* **39**: 123–128.

Jordan, C.F. and Escalante, G. 1980. Root productivity in an Amazonian rain forest. *Ecology* **61**: 14–18.

Jordan, C., Golley, F., Hall, J. and Hall, J. 1980. Nutrient scavenging of rainfall by the canopy of an Amazonian rain forest. *Biotropica* **12**: 61–66.

Jordan, C.F. and Herrera, R. 1981. Tropical rain forests: are nutrients really critical? *American Naturalist* **117**: 167–180.

Jordan, C.F. and Heuveldop, J. 1981. The water budget of an Amazonian rain forest. *Acta Amazonica* **11**: 87–92.

Jordan, C.F., and Farnworth, E.G. 1982. Natural vs. plantation forests: a case study of land reclamation strategies for the humid tropics. *Environmental Management* **6**: 485–492.

Jordan, C.F., Caskey, W., Escalante, G. Herrera, R. Montagnini, F., Todd, R. and Uhl, C. 1982. The nitrogen cycle in a "terra firme" rainforest on oxisol in the Amazon Territory of Venezuela. *Plant and Soil* **67**: 325–332.

Jordan, C., Caskey, W., Escalante, G., Herrera, R. Montagnini, F., Todd, R. and Uhl, C. 1983. Nitrogen dynamics during conversion of primary rain forest to slash and burn agriculture. *Oikos* **40**: 131–139.

Kato, R., Tadaki, Y. and Ogawa, H. 1978. Plant biomass and growth increment studies in Pasoh forest. *Malaysian Nature Journal* **30**: 211–224.

Kay, D.E. 1973. Root crops. *The Tropical Products Institute Crop and Product Digest.* No. 2.

Kellog, C.E. 1963. Shifting cultivation. *Soil Science* **95**: 221–230.

Kenworthy, J.B. 1971. Water and nutrient cycling in a tropical rain forest. Pages 49–59 in J.R. Flenly, ed. *The water relations of Malesian forests.* Transactions of the first Aberdeen–Hull symposium on Malesian Ecology. Institute for Southeast Asian Biology, University of Aberdeen, Aberdeen.

Kinkead, G. 1981. Trouble in D.K. Ludwig's jungle. *Fortune,* April 20, 1981: 102–117.

Kirkland, G.L. Jr. 1977. Responses of small mammals to the clearcutting of northern Appalachian forests. *Journal Mammalogy* **58**: 600–609.

Kline, J.R., and Jordan, C.F. 1968. Tritium movement in soil of tropical rain forest. *Science* **160**: 550–551.

Kline, J.R., Martin, J.R., Jordan, C.F. and Koranda, J.J. 1970. Measurement of transpiration in tropical trees with tritiated water. *Ecology* **51**: 1068–1073.

Kline, J.R., Reed, K.L., Waring, R.H. and Stewart, M.L. 1976. Field measurement of transpiration in Douglas fir. *Journal of Applied Ecology* **13**: 272–283.

Klinge, H. 1967. Podzol soils: a source of blackwater rivers in Amazonia. *Atas do Simposio sobre a Biota Amazonica* **3**: 117–125.

Klinge, H. 1978. Studies on the ecology of Amazon caatinga forest in southern Venezuela. *Acta Cientifica Venezolana* **29**: 257–262.

Klinge, H. and Herrera, R. 1978. Biomass studies in Amazon caatinga forest in southern Venezucla. 1. Standing crop of composite root mass in selected stands. *Tropical Ecology* **19**: 93–110.

Klinge, H. and Medina, E. 1978. Rio negro caatingas and campinas, Amazonas states of Venezuela and Brazil. Pages 483–488 in R.L. Specht ed. *Ecosystems of the World,* 9A. Heathlands and related shrublands. Elsevier, Amsterdam.

Kronberg, B.I., Fyfe, W.S., Leonardos, O.H. and Santos, A.M. 1979. The chemistry of some Brazilian soils: element mobility during intense weathering. *Chemical Geology* **24**: 211–229.

Kronberg, B.I., Fyfe, W.S., McKinnnon, B.J., Couston, J.F., Stilianidi-Filho, B. and Nash, R.A. 1982. Model for bauxite formation: Paragominas (Brazil). *Chemical Geology* **35**: 311–320.

Lal, R. 1981. Deforestation of tropical rainforest and hydrological problems. Pages 131–140 in R. Lal and E.W. Russcll cds. *Tropical agricultural hydrology: watershed management and land use.* Wiley, New York.

Lavelle, P. 1978. Les ver de terre de la savane de Lamto (Cote d'Ivoire). Peuplements, populations et fonctions dan l'ecosteme. Ecole Normale Superieure. *Pub. Lab. Zool.* No. 12. Paris.

Lavelle, P., Maury, M.E. and Serrano, V. 1981. Estudio cuantitativo de la fauna del suelo en la region de Laguna Verde, Vera Cruz. epoca de lluvias. *Inst. Ecol. Pub.* **6**: 75–105.

Lemee, G. Undated. Recherches sur l'ecosysteme de la foret subequatoriale de basse Cote-d'Ivoire. Unpublished.

Likens, G.E. and Eaton, J.S. 1970. A polyurethane stemflow collector for trees and shrubs. *Ecology* **51**: 938–939.

Lieth, H. 1975. Measurement of caloric values. Pages 119–129 in H. Lieth and R.H. Whittaker, eds. *Primary productivity of the biosphere.* Springer-Verlag, New York.

Likens, G.E., Bormann, F.H., Pierce, R.S., Eaton, J.S. and Johnson, N.M. 1977. *Biogeochemistry of a forested ecosystem.* Springer-Verlag, New York.

Lim, M.T. 1978. Litterfall and mineral nutrient content of litter in Pasoh Forest Reserve. *Malaysian Nature Journal* **30**: 375–380.

Lizot, J. 1977. Population, resources, and warfare among the Yanomami. *Man* **12**: 497–517.

Ljungstrom, P.O. and Reinicke, A.J. 1969. Ecology and natural history of microchaetid earthworms of South Africa. *Pedobiologia* **9**: 152–157.

Loveless, A.R. 1961. A nutritional interpretation of sclerophylly based on differences in the chemical composition of sclerophyllous and mesophytic leaves. *Annals Botany* **25**: 168–184.

Loveless, A.R. 1962. Further evidence to support a nutritional interpretation of sclerophylly. *Annals Botany* **26**: 551–561.

Lowe-McConnell, R.H. 1975. *Fish communities in tropical freshwaters: their distribution, ecology, and evolution.* Longman. London.

Lowman, M.D., and Box, J.D. 1983. Variation in leaf toughness and phenolic content among five species of Australian rain forest trees. *Australian Journal of Ecology* **8**: 17–25.

Lozano, J.C., Belloti, A., van Schoonhoven, A., Howeler, R., Doll, J., Howell, D. and Bates, T. 1976. *Field problems in cassava.* CIAT, Cali, Colombia.

Luc, M. 1968. Nematological problems in the former French African tropical territories and Madagascar. Pages 93–112 in G.C. Smart and V.G. Perry, eds. *Tropical Nematology.* University of Florida Press, Gainesville.

Luse, R.A. 1970. The phosphorus cycle in a tropical rain forest. Pages H-161–H-166 in H.T. Odum ed. *A tropical rain forest.* Division of Technical Information, U.S. Atomic Energy Commission, Washington, D.C.

Luvall, J.C. 1984. *Tropical deforestation and recovery: the effects on the evapotranspiration process.* Ph.D. Dissertation, University of Georgia, Athens.

Maass, M. Energy investment strategy in oligotrophic tropical ecosystems. 1982 term paper, Univ. of Georgia. Unpublished.

Madge, D.S. 1969. Field and laboratory studies on the activities of two species of tropical earthworms. *Pedobiologia* **9**: 188–214.

Matsumoto, T. 1976. The role of termites in an equatorial rain forest ecosystem of west Malaysia. *Oecologia* **22**: 153–178.

McBrayer, J.F., Ferris, J.M., Metz, L.J., Gist, C.S., Cornaby, B.W., Kitazawa, Y., Kitazawa, L., Wernz, J.G., Krantz, G.W. and Jensen, H. 1977. Decomposer invertebrate populations in U.S. forest biomes. *Pedobiologia* **17**: 89–96.

McGill, W.B. and Cole, C.V. 1981. Comparative aspects of cycling of organic C,N,S and P through soil organic matter. *Geoderma* **26**: 267–286.

McGill, W.B. and Christie, E.K. 1983. Biogeochemical aspects of nutrient cycle interactions in soils and organisms. Pages 271–301 in B. Bolin and R.B. Cook eds. *The major biogeochemical cycles and their interactions.* Wiley. Chichester.

Medina, E. and Zelwer, M. 1972. Soil respiration in tropical plant communities. Pages 245–267 in P.M. Golley and F.B. Golley eds. *Tropical ecology with an emphasis on organic productivity.* Institute of Ecology, University of Georgia, Athens.

Medina, E., Herrera, R., Jordan, C. and Klinge, H. 1977. The Amazon project of the Venezuelan Institute for Scientific Research. *Nature and Resources* (UNESCO) 8, No. 3: 4–6.

Medina, E., Sobrado, M. and Herrera, R. 1978. Significance of leaf orientation for leaf temperature in an Amazonian sclerophyll vegetation. *Radiation and Environmental Biophysics* **15**: 131–140.

Medina, E., Klinge, H., Jordan, C. and Herrera, R. 1980. Soil respiration in Amazonian rain forests in the Rio Negro Basin. *Flora* **170**: 240–250.

Medina, E. and Klinge, H. 1983. Productivity of tropical forests and tropical woodlands. Pages 281–303 in O.L. Lange, P.S. Nobel, C.B. Osmond, and H. Ziegler, eds. *Physiological plant ecology* IV. Ecosystem processes: mineral cycling, productivity and man's influence. Springer-Verlag, Berlin.

Meentemeyer, V. *AET values for selected ecological research sites.* Geography Dept.,

Univ. of Georgia, Athens, GA.

Meggars, B.J. 1971. *Amazonia: man and culture in a counterfeit paradise*. Aldine-Atherton. Chicago.

Meyer, F.H. and Gottsche, D. 1971. Distribution of root tips and tender roots of beech. Pages 48–52 in H. Ellenberg ed. Integrated experimental ecology. Methods and results of ecosystem research in the German Solling project. *Ecological Studies*, 2. Springer-Verlag, Heidelberg.

Montagnini, F. and Jordan, C.F. 1983. The role of insects in the productivity of cassava (Manihot esculenta Crantz) on a slash and burn site in the Amazon Territory of Venezuela. *Agriculture, Ecosystems and Environment*, 9: 293–301.

Monteith, J.L. 1965. Evaporation and environment. The state and movement of water in living organisms. Pages 205–234 in *19th Symposium Society for Experimental Biology*.

Monteith, J.L. 1973. *Principles of environmental physics*. Edward Arnold. London.

Nabholz, J.V. 1973. *Small mammals and mineral cycling on three Coweeta watersheds*. M.Sc. Thesis, University of Georgia, Athens, Georgia.

Nemeth, A. and Herrera, R. 1982. Earthworm populations in a Venezuelan tropical rain forest. *Pedobiologia* 23: 437–443.

Norman, M.J.T. 1979. *Annual cropping systems in the tropics*. University of Florida Press. Gainesville, Florida.

Normanha, E.S. 1970. General aspects of cassava root production in Brazil. Pages 61–63 in *Proceedings Second International Symposium on Tropical Root Crops*. Honolulu, Hawaii.

Nye, P.H. and Greenland, D.J. 1960. *The soil under shifting cultivation*. Technical Comm. No. 51, Commonwealth Bureau of Soils, Harpenden, Commonwealth Agricultural Bureaux, Farnham Royal, Bucks, England.

Nye, P.H. and Greenland, D.J. 1964. Changes in the soil after clearing tropical forest. *Plant and Soil* 21: 101–112.

Odum, E.P. 1969. The strategy of ecosystem development. *Science* 164: 262–270.

Odum, E.P. 1985. Trends expected in stressed ecosystems. *BioScience* 35: 419–422.

Odum, E.P., Finn, J.T. and Franz, E.H. 1979. Perturbation theory and the subsidy-stress gradient. *BioScience* 29: 349–352.

Odum, H.T. 1970a. Summary: an emerging view of the ecological system at El Verde. Pages I-191–I-289 in H.T. Odum ed. *A tropical rain forest*. Div. Technical Information. U.S. Atomic Energy Commission. Washington, D.C.

Odum, H.T. 1970b. Rain forest structure and mineral cycling homeostasis. Pages H-3–H-52 in H.T. Odum ed. *A tropical rain forest*. Div. Technical Information. U.S. Atomic Energy Commission. Washington, D.C.

Ogawa, H. 1978. Litter production and carbon cycling in Pasoh forest. *Malaysian Nature Journal* 30: 367–373.

Olsen, S.R. and Dean, L.A. 1965. Phosphorus. Pages 1035–1049 in C.A. Black ed. *Methods of soil analysis*, part 2, No. 9 in the series Agronomy. American Society of Agronomy, Madison, Wisconsin.

Olson, J.S. 1963. Energy storage and the balance of producers and decomposers in ecological systems. *Ecology* 44: 322–332.

Olson, J.S., Watts, J.A. and Allison, L.J. 1983. *Carbon in live vegetation of major world ecosystems*. Environmental Sciences Division Pub. No. 1997, Oak Ridge National Laboratory, Oak Ridge, Tenn.

Orians, G.H. 1969. The number of bird species in some tropical forests. *Ecology* 50: 783–801.

Parker, G.G. 1983. Throughfall and stemflow in the forest nutrient cycle. *Advances in ecological research* 13: 57–133. Academic Press, London.

Petriceks, J. 1968. *Shifting cultivation in Venezuela*. Ph.D. dissertation. SUNY College of Forestry, Syracuse, N.Y.

Pimentel, D., Hurd, L.E., Bellotti, A.C., Forster, M.J., Oka, I.N., Sholes, O.D. and Whitman, R.J. 1973. Food production and the energy crisis. *Science* **182**: 443–449.

Powell, P.E., Cline, G.R., Reid, C.P.P. and Szaniszlo, P.J. 1980. Occurrence of hydroxamate siderophore iron chelators in soils. *Nature* **287**: 833–834.

Price, P.W. 1975. *Insect ecology*. Wiley, N.Y.

Proctor, J. 1983. Mineral nutrients in tropical forests. *Progress in Physical Geography* **7**: 422–431.

Putz, F.E. 1983. Liana biomass and leaf area of a "tierra firme" forest in the Rio Negro basin, Venezuela. *Biotropica* **15**: 185–189.

Putzer, H. 1984. The geological evolution of the Amazon basin and its mineral resources. Pages 15–46 in H. Sioli ed. *The Amazon, Limnology and landscape ecology of a mighty tropical river and its basin*. Junk, Dordrecht.

Rapport, D.J., Regier, H.A. and Hutchinson, T.C. 1985. Ecosystem behavior under stress. *American Naturalist* **125**: 617–640.

Raich, J.W. 1980. *Carbon budget of a tropical soil under mature wet forest and young vegetation*. M.S. Thesis, University of Florida, Gainesville.

Richards, P.W. 1952. *The tropical rain forest*. Cambridge University Press, Cambridge.

Richardson, C.J. and Lund, J.A. 1975. Effects of clear cutting on nutrient losses in aspen forests on three soil types in Michigan. Pages 673–686 in F.G. Howell, J.B. Gentry and M.H. Smith, eds. *Mineral cycling in southeastern ecosystems*. Energy Research and Development Administration, Washington, D.C.

Roberts, T. 1973. Ecology of fishes in the Amazon and Congo basins. Pages 239–254 in B.J. Meggers, E.S. Ayensu, and D.W. Duckworth, eds. *Tropical forest ecosystems in Africa and South America: a comparative review*. Smithsonian Institution Press. Washington, D.C.

Robinson, J.G. and K.H. Redford. 1986. Body size, diet, and population density of neotropical forest mammals. *American Naturalist* **128**: 665–680.

Saldarriaga, J., 1985. Ph.D. Dissertation. Univ. of Tennessee. Knoxville. In Prep.

Salick, J., Herrera, R. and Jordan, C.F. 1983. Termitaria: Nutrient patchiness in nutrient-deficient rain forests. *Biotropica* **15**: 1–7.

Sanchez, P.A. 1976. *Properties and management of soils in the tropics*. Wiley, New York.

Sanchez, P.A. 1981. Soils of the humid tropics. Pages 347–410 in *Blowing in the wind: deforestation and long-range implications*. Studies in third world societies. Dept. of Anthropology, College of William and Mary, Williamsburg, Virginia.

Sanchez, P.A., Bandy, D.E., Villachica, J.H. and Nicholaides, J.J. 1982. Amazon basin soils: management for continuous crop production. *Science* **216**: 821–827.

Sanchez, P.A., Villachica, J.H. and Bandy, D.E. 1983. Soil fertility dynamics after clearing a tropical rainforest in Peru. *Soil Science Society of America* **47**: 1171–1178.

Sandford, R.L. 1987. Apgeotropic roots in an Amazon Rain Forest. *Science* **235**: 1062–1064.

Sanford, R.L., Saldarriaga, J., Clark, K.E., Uhl, C. and Herrera, R. 1985. Amazon rain-forest fires. *Science* **227**: 51–53.

Shugart, H.H., Reichle, D.E., Edwards, N.T. and Kercher, J.R. 1976. A model of calcium cycling in an east-Tennessee Liriodendron forest: model structure parameters and frequency response analysis. Pages 445–456 in *l'Association Internationale des Sciences Hydrologiques Symposium de Tokyo*, Tokyo, Japan.

Singer, R. and I.J. Silva Araujo. 1979. Litter decomposition and ectomycorrhiza in Amazon forests. *Acta Amazonica* **9**: 25–41.

Sioli, H. 1973. *Recent human activities in the Brazilian Amazon region, and their ecological effects*. Pages 321–334 in B.J. Meggars, E.S. Ayensu, and W.D. Duckworth, eds. Smithsonian Institution Press. Washington, D.C.

Sioli, H. 1975. Tropical rivers as expressions of their terrestrial environments. Pages 275–288 in F.B. Golley and E. Medina eds. *Tropical ecological systems.* Springer Verlag. N.Y.

Smith, M.H., Gentry, J.B. and Pinder, J. 1974. Annual fluctuations in small mammal populations in an eastern hardwood forest. *Journal Mammalogy* **55:** 231–234.

Smith, M.S., Firestone, M.K. and Tiedje, J.M. 1978. The acetylene inhibition method for short-term measurement of soil denitrification and its evaluation using nitrogen-13. *Soil Science Society of America J.* **42:** 611–615.

Smith, M.S. and Tiedje, J.M. 1979. Phases of denitrification following oxygen depletion in soil. *Soil Biology Biochemistry* **11:** 261–267.

Smith, N.J.H. 1979. *A pesca no Rio Amazonas.* Conselho Nacional de Desenvolvimento Cientifico e Tecnologico. Instituto Nacional de Pesquisa da Amazonia, Manaus, Brasil. Falangola, Belem, Brasil.

Sobrado, M.A. and Medina, E. 1980. General morphology, anatomical structure, and nutrient content of sclerophyllous leaves of the "bana" vegetation of Amazonas. *Oecologia* **45:** 341–345.

Sollins, P., Grier, C.C., McCorison, F.M., Cromack, K. and Fogel, R. 1980. The internal element cycles of an old-growth Douglas fir ecosystem in western Oregon. *Ecological Monographs* **50:** 261–285.

Sollins, P., Cromack, K., Fogel, R. and Yan Li, C. 1981. Role of low-molecular weight organic acids in the inorganic nutrition of fungi and higher plants. Pages 607–620 in D.T. Wicklow and G.C. Carrol, eds. *The fungal community, its organization and role in the ecosystem.* Marcel Dekker, New York.

Souza-Serrao, E.A., Falesi, I.C., de Veiga, J.B. and Teixeira Neto, J.F. 1978. Productivity of cultivated pastures on low fertility soils in the Amazon of Brazil. Pages 195–225 in P.A. Sanchez and L.E. Tergas eds. *Pasture production in acid soils of the tropics.* Proceedings of a seminar held at CIAT, Cali, Colombia, 17–21 April, 1978.

Sprick, E.G. 1979. *Composicion mineral y contenido de fenoles foliares de especies lenosas de tres bosques contrastantes de la region Amazonica.* Thesis for the degree Licenciado en Biologia, Universidad Central de Venezuela, Facultad de Ciencias, Escuela de Biologia.

St. John, T.V. 1980. A survey of mycorrhizal infections in an Amazonian rain forest. *Acta Amazonica* **10:** 527–533.

St. John, T.V., and Anderson, A.B. 1982. A re-examination of plant phenolics as a source of tropical black water rivers. *Tropical Ecology* **23:** 151–154.

St. John, T.V. and Coleman, D.C. 1983. The role of mycorrhizae in plant ecology. *Canadian Journal of Botany* **61:** 1005–1014.

St. John, T.V. and Uhl, C. 1983. Mycorrhizae in the rain forest at San Carlos de Rio Negro, Venezuela. *Acta Cientifica Venezolano* **34:** 233–237.

Stark, N. and Spratt, M. 1977. Root biomass and nutrient storage in rain forest Oxisols near San Carlos de Rio Negro. *Tropical Ecology* **18:** 1–9.

Stark, N.M., and Jordan, C.F. 1978. Nutrient retention by the root mat of an Amazonian rain forest. *Ecology* **59:** 434–437.

Stevens, P.R. and Walker, T.W. 1970. The chronosequence concept and soil formation. *Quaterly Review of Biology* **45:** 333–350.

Stevenson, I.L. 1964. Biochemistry of soil. Pages 242–291 in F.E. Bear ed. *Chemistry of the soil.* Reinhold, New York.

Struthers, P.H. and Sieling, D.H. 1950. Effect of organic anions on phosphate precipitation by iron and aluminum as influenced by pH. *Soil Science* **69:** 205–213.

Swank, W.T. and Douglass, J.E. 1977. Nutrient budgets for undisturbed and manipulated hardwood forest ecosystems in the mountains of North Carolina. Pages 343–364 in D.L. Correll, ed. *Watershed research in eastern North America.* Chesapeake Bay Center for Environmental Studies, Edgewater, Maryland.

Swenson, R.M., Cole, C.V. and Sieling, D.H. 1949. Fixation of phosphate by iron and aluminum and replacement by organic and inorganic ions. *Soil Science* **67**: 3–22.

Swift, M.J. 1984. Soil biological processes and tropical soil fertility. Special Issue -5, *Biology International.* The International Union of Biological Sciences News Magazine.

Swift, M.J., Heal, O.W. and Anderson, J.M. 1979. Decomposition in terrerstrial ecosystems. *Studies in Ecology* 5. Univ. of California Press. Berkeley.

Technicon Industrial Systems. 1977. *Technicon industrial method* No. 369–75 A/B. Digestion and sample preparation for the analysis of total Kjeldahl nitrogen and/or total phosphorus in food and agricultural products using the Technicon BD-20 block digestor. Technicon Industrial Systems, Tarrytown, New York.

Time, 1976. Ludwig's wild Amazon kingdom. *Time Magazine,* Nov. 15, 1976: 59–59A.

Time, 1979. Billionare Ludwig's Brazilian gamble. *Time Magazine,* Sept. 10, 1979: 76–78.

Time, 1982. End of a billion-dollar dream. *Time Magazine,* Jan. 25, 1982: 59.

Todd, R.L., Meyer, R.D. and Waide, J.B. 1978. Nitrogen fixation in a deciduous forest in the southeastern United States. In U. Granhall ed. Environmental role of nitrogen-fixing blue-green algae and asymbiotic bacteria. *Ecological Bulletins* (Stockholm) **26:** 172–177.

Tosi, J.A. 1982. *Sustained yield management of natural forests.* Consultant Report. Tropical Science Center, San Jose, Costa Rica.

Turvey, N.D. 1974. Water in the nutrient cycle of a Papuan rain forest. *Nature* **251:** 414–415.

Uehara, G. and Gillman, G. 1981. *The mineralogy, chemistry and physics of tropical soils with variable charge clays.* Westview Press. Boulder, Colorado.

Uhl, C. 1980. *Studies of forest, agricultural and successional environments in the upper Rio Negro region of the Amazon Basin.* Ph.D. dissertation, Dept. of Botany and Plant Pathology, Michigan State University, East Lansing, Michigan.

Uhl, C. 1982. Tree dynamics in a species rich tierra firme forest in Amazonia, Venezuela. *Acta Científica Venezolana* **33:** 72–77.

Uhl, C. 1987. Factors controlling succession following slash-and-burn agriculture in Amazonia. *Journal of Ecology* **75**: 377–407.

Uhl, C. and Murphy, P.G. 1981a. Composition, structure and regeneration of a tierra firme forest in the Amazon Basin of Venezuela. *Tropical Ecology* **22:** 219–237.

Uhl, C. and Murphy, P.G. 1981b. A comparison of productivities and energy values between slash and burn agriculture and secondary succession in the upper Rio Negro region of the Amazon Basin. *Agro-Ecosystems* **7:** 63–83.

Uhl, C., Clark, K., Clark, H. and Murphy, P. 1981. Early plant succession after cutting and burning in the upper Rio Negro region of the Amazon Basin. *Journal of Ecology* **69:** 631–649.

Uhl, C., Clark, H. and Clark, K. 1982. Successional patterns associated with slash-and-burn agriculture in the upper Rio Negro region of the Amazon Basin. *Biotropica* **14:** 249–254.

Uhl, C. and Jordan, C.F. 1984. Succession and nutrient dynamics following forest cutting and burning in Amazonia. *Ecology* **65:** 1476–1490.

Vitousek, P. 1982. Nutrient cycling and nutrient use efficiency. *American Naturalist* **119:** 553–572.

Vitousek, P.M. 1983. The effects of deforestation on air, soil and water. Pages 223–

245 in B. Bolin and R.B. Cook eds. *The major biogeochemical cycles and their interactions.* Wiley. Chichester.

Vitousek, P. 1984. Litterfall, nutrient cycling, and nutrient limitations in tropical forests. *Ecology* **65**: 285–298.

Vitousek, P.M. and Reiners, W.A. 1975. Ecosystem succession and nutrient retention: a hypothesis. *BioScience* **25**: 376–381.

Vuilleumier, F. 1978. Remarques sur l' echantillonnage d'une riche avifaune de l'ouest de l'Ecuador. *L' Oiseau et R.F.O.* **48**: 21–36.

Wadsworth, F.H. 1981. Management of forest lands in the humid tropics under sound ecological principles. Pages 168–180 in F. Mergen, ed. *International symposium on tropical forest utilization and conservation.* Yale School of Forestry, New Haven, Connecticut.

Walker, T.W. and Adams, A.F.R. 1958. Studies on soil organic matter: I. Influence of phosphorus content of parent materials on accumulations of carbon, nitrogen, sulfur and organic phosphorus in grassland soils. *Soil Science* **85**: 307–318.

Walker, T.W. and Adams, A.F.R. 1959. Studies on soil organic matter: 2. Influence of increased leaching at various stages of weathering on levels of carbon, nitrogen, sulfur and organic and total phosphorus. *Soil Science* **87**: 1–10.

Walter, H. 1971. *Ecology of tropical and subtropical vegetation.* Oliver and Boyd. Edinburgh.

Watters, R.F. 1971. *Shifting cultivation in Latin America.* FAO Forestry Development Paper No. 17. Food and Agriculture Organization, Rome.

Went, F.W. and Stark, N. 1968a. Mycorrhiza. *BioScience* **18**: 1035–1039.

Went, F.W. and Stark, N. 1968b. The biological and mechanical role of soil fungi. *Proceedings of the National Academy of Sciences* **60**: 497–504.

Werner, D., Flowers, N.M., Ritter, M.L. and Gross, D.R. 1979. Subsistence productivity and hunting effort in native South America. *Human Ecology* **7**: 303–315.

Whitford, W.G. 1976. Temporal fluctuations in density and diversity of desert rodent populations. *Journal Mammology* **57**: 351–369.

Whitmore, T.C. 1984. *Tropical rain forests of the far East.* Clarendon Press, Oxford.

Whittaker, R.H. and Likens, G.E. 1975. The biosphere and man. Pages 305–328 in H. Lieth and R.H. Whittaker eds. *Primary productivity of the biosphere.* Springer Verlag. N.Y.

Wiegert, R.G. 1970. Effects of ionizing radiation on leaf fall, decomposition and litter microarthropods of a montane rain forest. Pages H-89–H-100 in H.T. Odum ed. *A tropical rain forest.* U.S. Atomic Energy Commission. Washington, D.C.

Witkamp, M. 1970. Mineral retention by epiphyllic organisms. Pages H-177–H-179 in H.T. Odum ed. *A tropical rain forest.* Division of Technical Information, U.S. Atomic Energy Commission, Washington, D.C.

Wood, T.G. and Sands, W.A. 1978. The role of termites in ecosystems. Pages 245–292 in M.V. Brian ed. *Production ecology of ants and termites.* Cambridge Univ. Press. Cambridge.

Woodwell, G.M. 1970. Effects of pollution on the structure and physiology of ecosystems. *Science* **168**: 429–433.

Woodwell, G.M., Whittaker, R.H. and Houghton, R.A. 1975. Nutrient concentrations in plants in the Brookhaven oak-pine forest. *Ecology* **56**: 318–332.

Woodwell, G.M. and Whittaker, R.H. 1967. Primary production and the cation budget of the Brookhaven forest. Pages 151–166 in H.E. Young, ed. *Symposium on primary productivity and mineral cycling in natural ecosystems.* College of Life Sciences and Agriculture, University of Maine, Orono.

Wright, R.F. 1976. The impact of forest fire on the nutrient fluxes to small lakes in northeastern Minnesota. *Ecology* **57**: 649–663.

Yoda, K. 1978. Organic carbon, nitrogen, and mineral nutrients stock in the soils of Pasoh forest. *Malaysian Nature Journal* **30:** 229–251.

Zinke, P.J., Sabhasri, S. and Kunstadter, P. 1978. *Soil fertility aspects of the Lua' forest fallow system of shifting cultivation.* Pages 134–159 in P. Kunstadter, E.C. Chapman, and S.S. Sabhasri, eds. University of Hawaii Press. Honolulu.

INDEX

Fertilizers, 48, 88, 106, 108
Fish, 36
    methods of community studies, 125
    population density, 36, 38, 39
    species diversity, 38
    species list, 124
Forest biomass, 20–25
    determination of, above ground, 133
        above-ground variability, 135
        comparisons of above and below
            ground dry weight biomass, 137
        forest floor and below ground, 135
        remains of forest in cultivated plot, 136
        total biomass variability, 136
    productivity, determination of
        above-ground biomass increment, 138,
            139
        litter fall, 140
        mortality, and dynamics of biomass
            stocks, 138
        root productivity, 140
    site locations used for comparisons, *24*

Geology, 7–9, 66, 107
Guiana shield, 8, 9

Herbivory, 18, 56, 73
Humic acids, 18
Humus, 2, *16*
Humus mat, 10, 31, 34, 50, 52
    cycling index, 31

Igapo, 14
Insect consumption, 18, 56, 89
Inselbergs, 9

Land use capability, 66
Laterization process, 9, 10, 12
Latosols, 83, 84
Leachate nutrient concentration (see under
    nutrient leaching)
Leaf, abscission and nutrient
    concentrations, 57, 59
    area index, 64
    biomass, 48, 62, 64
    caloric values, 56–58
    cutter ants, 42
    litter decay constant, 64
    litter nutrient-use efficiency, 48, 49, 64
    litter production, 64
    production, 64
    turn-over time, 56, 64
Leaves, scleromorphic, 56, 64, 65
    and throughfall nutrient concentrations,
        56, 58
Life-span, of crop species, 99
    of native successional vegetation, 99
Location of study, *5, 6*

Lysimeter, 97

Magnesium, cycling model, 33
    in humus mat, 34, 98
    nutrient leaching, 97
    stock and cumulative losses, 80, 81
Mammals, 36
    density, 36, 37
    methods of community studies, 125–126
    population density, 36
    species list, 120–122
Management strategies, 103–109
    economic disadvantages of selective and
        strip-harvesting, 105, 106
    selective harvesting, 105
    silviculture, 105
    strip-harvesting, 105–108
Microbial carbon biomass, 45, 46
    growth, 87, 89
Models, of calcium cycle, 31
    of magnesium cycle, 33
    of potassium cycle, 32
Monoculture plantations, 27, 107
Mycorrhizae, 53–56, 100, 106
    and phosphate uptake, 53
    ectomycorrhizae, 53
    oxalic acid productivity by, 88
    vesicular-arbuscular, endomycorrhizae, 53

Native successional vegetation, 98–102
Net biomass production, *26*
Net primary productivity, as symptom of
    stress, 25
    following abandonment, 93, 95, 96
    of control forest, 73, 74
    of crops, edible, 72, 74
    of crops, total, 72, 74
    of slash-and-burn plot, 69, 71, 74
Nitrogen, as limiting nutrient, 48, 64
    availability, 67, 88
    dynamics, 46, 47, 88
    fixation, 46, 57, 87
    flux, 97
    inhibition by phosphorus lack, 87, 88
    interactions with phosphorus, 87
    stock and cumulative losses, 80, 81
Nutrient balance, 27, 71
    concentration, 99, 101
        after cut-and-burn, 77, 83
        in biomass, determination of in forest,
            142, 143
            crop and successional vegetation,
            143
        in secondary successional vegetation,
            99, 101
    conservation mechanisms, 57, 59, 69
    cycling, 29–34
        comparisons for calcium and